An Introduction to Geological Structures and Maps

An Introduction to Geological Structures and Maps

Fifth Edition

G. M. Bennison

Formerly Senior Lecturer in Geology, University of Birmingham

Edward Arnold

A division of Hodder & Stoughton

LONDON   NEW YORK   MELBOURNE   AUCKLAND

© 1990 G. M. Bennison

First published in Great Britain 1964
Reprinted 1965, 1966, 1968
Second edition 1969
Reprinted 1971, 1973
Third edition 1975
Reprinted 1977, 1979, 1981, 1982
Fourth edition 1985
Reprinted 1985, 1986, 1987, 1989
Fifth edition 1990

Distributed in the USA by Routledge, Chapman and Hall, Inc.,
29 West 35th Street, New York, NY 10001

British Library Cataloguing in Publication Data

Bennison, G. M. (George Mills), *1922–*
    An introduction to geological structures and maps. – 5th ed.
    1. Geological maps. Map reading
    I. Title
    551.8

    ISBN 0-340-51760-3

Typeset in 10/11 pt Helios Light by Colset Private Limited, Singapore
Printed and bound in Great Britain for Edward Arnold, the
educational, academic and medical publishing division of Hodder
and Stoughton Limited, Mill Road, Sevenoaks, Kent TN13 2YD by
The Bath Press and W.H. Ware, Avon

Contents

Key to maps   vi

Preface   vii

1 **Horizontal and dipping strata**   1
Contours, section drawing, vertical exaggeration; Dip, structure contours, construction of structure contours, section drawing; True and apparent dip; Calculation of thickness of a bed, vertical and true thickness; Width of outcrop; Inliers and outliers.

2 **'Three-point' problems**   11
Construction of structure contours; Insertion of outcrops; Depth in boreholes.

3 **Unconformities**   16
Overstep; Overlap; Sub-unconformity outcrops.

4 **Folding**   19
Anticlines and synclines; Asymmetrical folds, overfolds and isoclinal folds; Similar and concentric folding; Two possible directions of strike.

5 **Faults**   27
Normal and reversed faults; The effects of faulting on outcrops; Classification of faults, dip-slip, strike-slip and oblique-slip faults; Dip faults and strike faults; Calculation of the throw of a fault; Wrench or tear faults; Pre- and post-unconformity faulting; Structural inliers and outliers; Posthumous faulting; Isopachytes.

6 **More folds and faulted folds**   40
Plunging folds, calculation of the amount of plunge; The effect of faulting on fold structures; Displacement of folds by strike-slip (wrench) faults, calculation of strike-slip displacement; Faults parallel to the limbs of a fold; Sub-surface structures; Posthumous folding; Polyphase folding; Bed isopachytes.

7 **Complex structures**   53
Nappes; Thrust faults; Axial plane cleavage.

8 **Igneous rocks**   58
Concordant intrusions, sills; Lava flows and tuffs; Discordant intrusions, dykes, ring-dykes, cone-sheets, stocks, bosses and batholiths, volcanic necks.

Description of a geological map; Geological history of Map 28.   64

Numerical answers   66

Index   67

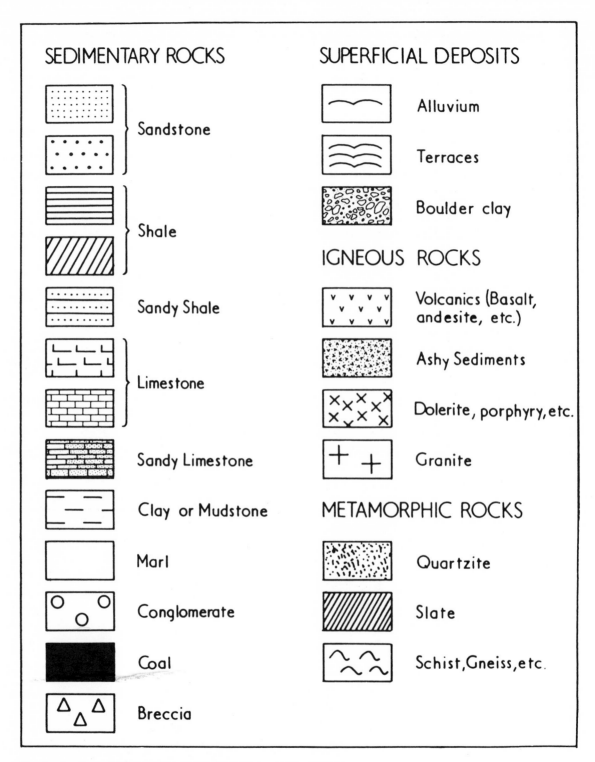

SEDIMENTARY ROCKS

Sandstone

Shale

Sandy Shale

Limestone

Sandy Limestone

Clay or Mudstone

Marl

Conglomerate

Coal

Breccia

SUPERFICIAL DEPOSITS

Alluvium

Terraces

Boulder clay

IGNEOUS ROCKS

Volcanics (Basalt, andesite, etc.)

Ashy Sediments

Dolerite, porphyry, etc.

Granite

METAMORPHIC ROCKS

Quartzite

Slate

Schist, Gneiss, etc.

Key to shading widely used on geological maps and text figures.

Preface to previous editions

This book is designed primarily for university and college students taking geology as an honours course or as a subsidiary subject. Its aim is to lead the student by easy stages from the simplest ideas on geological structures right through the first year course on geological mapping, and much that it contains should be of use to students of geology at GCE 'A' level. The approach is designed to help the student working with little or no supervision: each new topic is simply explained and illustrated by text-figures, and exercises are set on succeeding problem maps. If students are unable to complete the problems they should read on to obtain more specific instructions on how the theory may be used to solve the problem in question. Problems relating to certain published geological survey maps are given at the end of most chapters. Some of the early maps in the book are of necessity somewhat 'artificial' so that new structures can be introduced one at a time thus retaining clarity and simplicity. Structure contours (see p. 5) are seldom strictly parallel in nature; it is therefore preferable to draw them freehand, though – of course – as straight and parallel as the map permits. In all cases except the three-point problems, the student should examine the maps and attempt to deduce the geological structures from the disposition of the outcrops in relation to the topography, as far as this is possible, before commencing to draw structure contours.

The author wishes to thank Dr F. Moseley for making many valuable suggestions when reading the manuscript of this book, Dr R. Pickering and Dr A.E. Wright for their continuing help and interest.

Preface to this edition

Through successive editions additional material had been fitted into the existing 64 pages. Now an enlarged fifth edition provides the opportunity to add new maps and more explanatory text. The aim is to make the progression from the simplest map to the most difficult more gradual by increasing the number of steps. A further three maps (out of 5 new maps) are not dependent on structure contours for their solution. The point is re-emphasized that the student should always examine a map with a view to deducing the basic structures from outcrop patterns (in relation to the topography), as far as this is possible, even on maps where the precise solution is dependent on structure contours. The opportunity has been taken to update terminology, particularly of fold and fault structures. The order of presentation has been radically changed from earlier editions and it is hoped that this will be found to be more logical. In addition to problem maps based on, or adapted from, published geological maps, reference is made in each chapter to British Geological Survey maps (generally on the 1:50,000 scale). They are selected to illustrate the problems dealt with in each chapter. While it is appreciated that only some readers will have access to these maps, the necessity of relating problem maps to 'real' geological maps justifies the small amount of the book devoted to these exercises.

My thanks are due to friends and colleagues for their continuing interest and helpful suggestions. I am particularly grateful to Dr K.A. Moseley and Dr. D.E. Roberts. Without the expertise and diligence of Dr R. Pickering this edition would not have been possible. My thanks to Mr Carl Burness for drafting the new maps and diagrams and revising many of the figures.

Readers wishing to pursue the subject further are recommended the following reading:

Advanced Geological Map Interpretation, F. Moseley, Edward Arnold, 1979.
The use of Stereographic Projections in Structural Geology, F.C. Phillips, Edward Arnold, 1954.
The Mapping of Geological Structures, K. McClay, Geological Society of London Handbook, 1987.

Belbroughton   G.M. Bennison
June, 1989

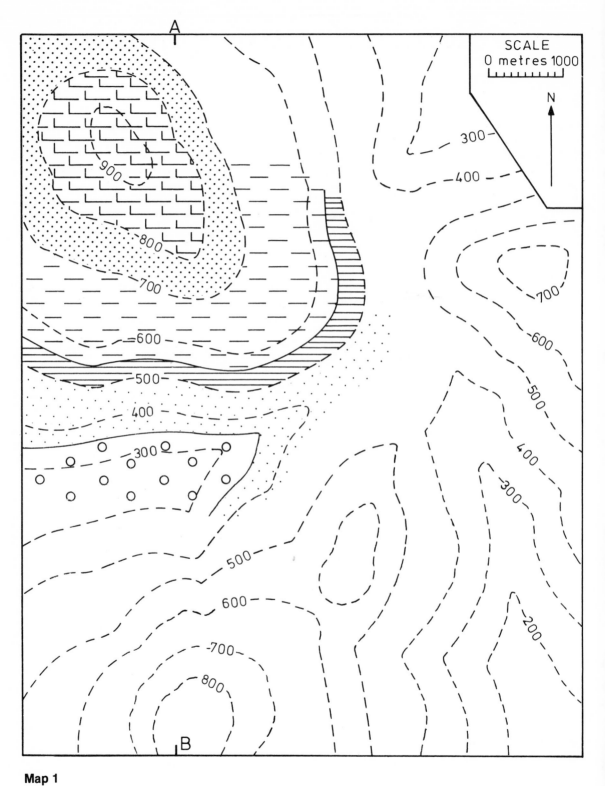

SCALE
0 metres 1000

N

300
400
700
600
500
400
300
200

900
800
700
600
500
400
300

A
B

500
600
700
800

Map 1

1   Horizontal and dipping strata

Contours

Hills and valleys are usually carved out of layered sequences of rock, or strata, the individual members – or beds – differing in thickness and in resistance to erosion. Hence diverse topography (surface features) and land-forms are produced. Only in exceptional circumstances is the topography eroded out of a single rock-type.

In the simplest case we can consider strata are horizontal. Rarely are they so in nature; they are frequently found elevated hundreds of metres above their position of deposition, and tilting and warping has usually accompanied such uplift. The pattern of outcrops of the beds where the strata are horizontal is a function of the topography; the highest beds in the sequence (the youngest) will outcrop on the highest ground and the lowest beds in the sequence (the oldest) will outcrop in the deepest valleys. Geological boundaries will be parallel to the contour lines shown on a topographic map for they are themselves contour lines, since a contour is a line joining an infinite number of points of the same height.

Section drawing   Draw a base line the exact length of the line A–B on Map 1 (19.0 cm). Mark off on the base line the points at which the contour lines cross the line of section: for example, 8.5 mm from A mark a point corresponding to the intersection of the 700 m contour. From the base line erect a perpendicular corresponding in length to the height of the ground and, since it is important to make vertical and horizontal scales equal wherever practicable, a perpendicular of length 14 mm must be erected to correspond to the 700 m contour (since 1000 m = 2 cm and 100 m = 2 mm) (Fig. 1). Sections can readily be drawn on metric squared paper (or on 1/10″ in some cases).

Map 2 shows an area of the Cotswold Hills and adjacent lowlands where the strata are virtually horizontal. Geological boundaries therefore are parallel to topographic contours, a point made on page 5. Contours have been omitted for clarity but the general heights of hills and plain are given. These, together with the altitude of a number of points (spot-heights and triangulation points) enable us to draw a sufficiently accurate topographic profile of a section across the map.

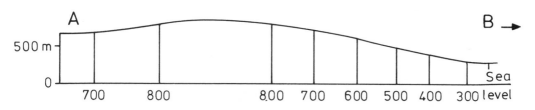

Fig. 1   Part of a section along the line A–B on Map 1 to show the method of drawing the ground surface (or profile).

Map 1.   The geological outcrops are shown in the north-west corner of the map. It can be seen that the beds are horizontal as the geological boundaries coincide with, or are parallel to, the ground contour lines. Complete the geological outcrops over the whole map. Indicate the position of a spring-line on the map. How thick is each bed? Draw a vertical column showing each bed to scale, 1 cm = 100 m. Draw a section along the line A–B. (Contours in metres.)

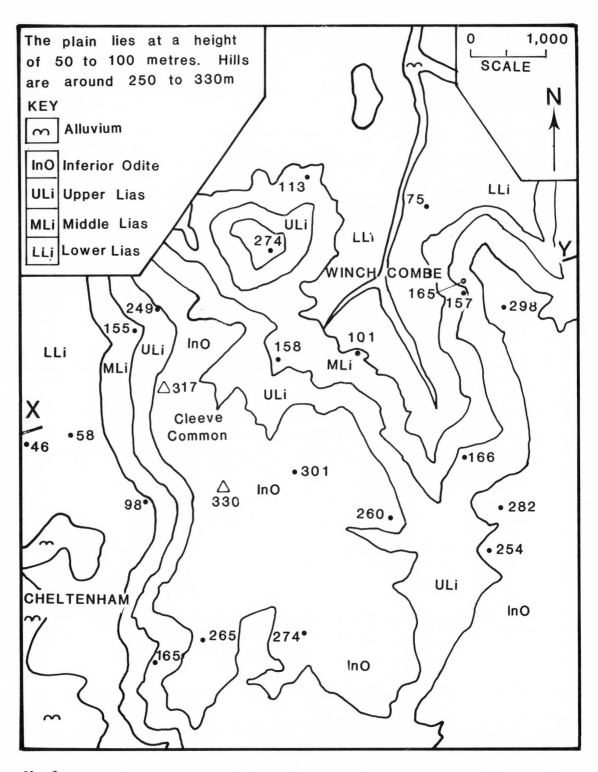

The plain lies at a height of 50 to 100 metres. Hills are around 250 to 330m

KEY

⌒	Alluvium
InO	Inferior Odite
ULi	Upper Lias
MLi	Middle Lias
LLi	Lower Lias

0 — 1,000
SCALE

N

113

75

ULi

LLi

274

WINCH COMBE

165 157

• 298

249

155

LLi

ULi

InO

158

101

MLi

Y

△ 317

MLi

ULi

X

Cleeve Common

• 166

•58

• 46

• 301

△ 330

InO

98 •

260 •

• 282

• 254

ULi

CHELTENHAM

InO

• 265

274 •

165

InO

Map 2

2

Vertical Exaggeration The horizontal scale of a map has been determined at the time of the original survey. Commonly, we shall find it given as 1:50,000 – 1cm on the map representing 50,000 cms or 500 metres on level ground (1:50,000 is the 'Representative Fraction'). Maps on the scale of 1:25,000 and 1:10,000 are common. On older maps one inch represented one mile (1:63,360) and on USA maps two miles to the inch is a usual scale.

If a similar scale to the horizontal is used vertically we shall often find that the section is difficult to draw and that it is very difficult to include the geological details. For example, on Map 2 the highest hill on the section at 317 m would be only just over 0.6 cms above the sea-level base line. A suggested vertical scale of 1 cm = 200 m, with horizontal scale predetermined as 1 cm = 500 m gives a vertical exaggeration of 500/200 or 5/2 or 2.5.

Complications arise where the strata are inclined and further considerations of vertical exaggeration are discussed on p. 10.

Dip

Inclined strata are said to be dipping. The angle of dip is the maximum angle measured between the strata and the horizontal (regardless of the slope of the ground) (Fig. 2).

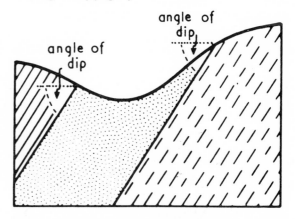

Fig. 2 Section showing dipping strata. The angle of dip is measured from the horizontal.

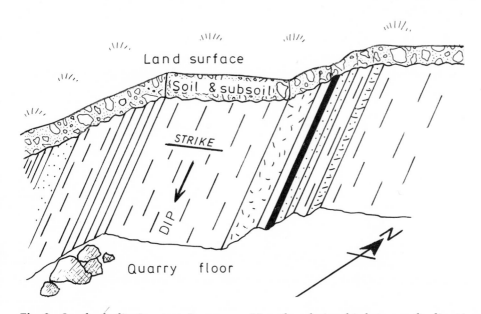

Fig. 3 Southerly dipping strata in a quarry. Note the relationship between the directions of dip and strike.

Map 2 This map is based on a portion of the British Geological Survey map of Moreton-in-the-Marsh, 1:50,000 scale, Sheet 217. (It is slightly simplified and reduced to fit the page size) Reproduced by permission of the Director, British Geological Survey: NERC copyright reserved. Draw a section along the line X–Y to illustrate the geology, using a vertical scale of 1 cm = 200 metres.

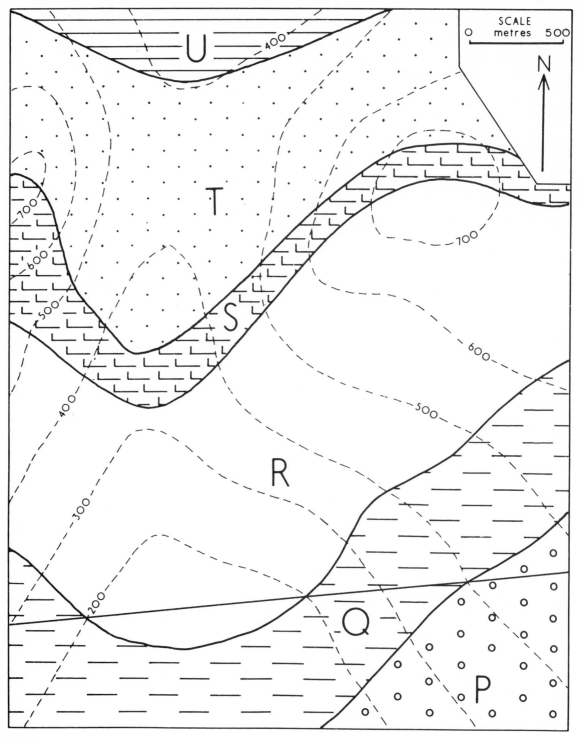

Map 3

The direction of dip is given as a compass bearing reading from 0 to 360°. For example, a typical reading would be 12/270. The first figure is the angle of dip, the angle the strata make with the horizontal. The second figure is the direction of that dip measured round from north in a clockwise direction (in this example due west). In North America, where the use of the Brunton Compass is almost universal, dip directions are more commonly given in relation to the main points of the compass, e.g. 10° west of south (i.e. 190°).

In a direction at right angles to the dip the strata are horizontal. This direction is called the strike (Fig. 3). An analogy may be made with the lid of a desk. A marble would roll down the desk lid in the direction of maximum dip. The edge of the desk lid, which is the same height above the floor along the whole of its length, i.e. it is horizontal, is the direction of strike.

Structure contours (= strike lines)

Just as it is possible to define the topography of the ground by means of contour lines, so we can draw contour lines on a bedding plane. These we call structure contours or strike lines, the former since they join points of equal height, the latter since they are parallel to the direction of strike. The terms are synonymous, but for the purposes of this book the term structure contour will be used.

Construction of structure contours

The height of a geological boundary is known where it crosses a topographic contour line. For example, the boundary between beds S and T on

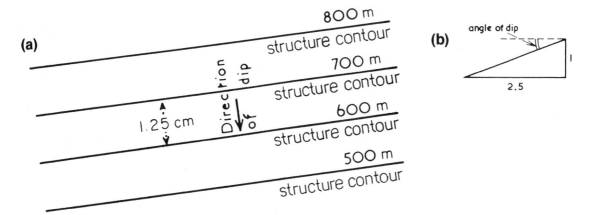

Fig. 4 (a) Plan showing structure contours and (b) section through contours showing the relationship between dip and gradient.

Map 3 The continuous lines are the geological boundaries separating the outcrops of the dipping strata, beds P, Q, R, S, T and U. Examine the map and note that the geological boundaries are not parallel to the contour lines but, in fact, intersect them. This shows that the beds are dipping. Before constructing structure contours can we deduce the direction of dip of the beds from the fact that their outcrops 'V' down the valley? Can we deduce the direction of dip if we are informed that Bed U is the oldest and Bed P is the youngest bed of the sequence? Draw structure contours for each geological interface* and calculate the direction and amount of dip. (Contours in metres.) Instructions for drawing structure contours are given below and one structure contour on the Q/R geological boundary has been inserted on the map as an example.

* Some confusion may arise since the term geological boundary is often applied both to the interface (or surface) between two beds also to the outcrop of that interface. It seems a satisfactory term to employ, however, since the two are related and the context generally avoids ambiguity.

map 3 cuts the 700 m contour at three points. These points lie on the 700 m structure contour which can be drawn through them. Since these early maps portray simply inclined plane surfaces, structure contours will be straight, parallel and – if dips are constant – equally spaced.

Having found the direction of strike 85°, we know that the direction of dip is at right angles to this, but we must ascertain whether the dip is 'northerly' or 'southerly'. A second structure contour can be drawn on the same geological boundary S–T through the two points where it cuts the 600 m contour. From the spacing of the structure contours we can calculate the dip or gradient of the beds (Fig. 4a).

Gradient = 700 m − 600 m in 1.25 cm
i.e. = 100 m in 1.25 cm.

As the scale of the map is given as 2.5 cm = 500 m, 100 m in 1.25 cm = 100 in 250 m.

Hence, **the gradient is 1 in 2.5, to 175°**

Frequently it is more convenient to utilize gradients, although on geological maps the dip is always given as an angle. By simple trigonometry we see that the angle of dip in the above case is that angle which has a tangent of $\frac{1}{2.5}$ or 0.4, i.e. 22°, to 175°. (Fig. 4b).

Section drawing The topographic profile is drawn by the method already described on page 1. The geological boundaries (interfaces) can be inserted in an analogous way by marking the points at which the line of section is cut by structure contours. Perpendiculars are then drawn from the base line, of length corresponding to the height of the structure contours (Fig. 5).

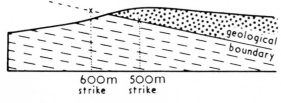

Fig. 5 Section to show the method of accurately inserting geological boundaries.

True and apparent dip

If the slope of a desk lid, or of a geological boundary (interface) or bedding plane, is measured in any direction between the strike direction and the direction of maximum dip, the angle of dip in that direction is known as an apparent dip (Fig. 6a). Its value will lie between 0° and the value of the maximum or true dip. Naturally occurring or man-made sections through geological strata (cliffs, quarry faces, road and rail cuttings) are unlikely to be parallel to the direction of true dip of the strata. What may be observed in these sections, therefore, is the dip of the strata in the direction of the section, i.e. an apparent dip (somewhat less than true dip in angle). The trigonometrical relationship is not simple:

Tangent apparent dip = Tangent true dip × Cosine β (see Fig. 6b; and Table II, p. 66).

However, the problem of apparent dip calculation is much simplified by considering it as a gradient. Just as the gradient of the bed in the direction of maximum dip is given by the spacing of the structure contours (1.9 cm = 380 m in Fig. 6b, representing a gradient of 1 in 3.8, since the scale of the map is 1 cm = 200 m), so the gradient in the direction in which we wish to obtain the apparent dip is given by the structure contour spacing measured in that direction (3.25 cm = 650 m in Fig. 6b, representing a gradient of 1 in 6.5).

In road and rail cuttings the direction of dip of strata is of vital importance to the stability of the slopes. Where practicable a cutting would be parallel to the direction of dip of the strata, minimizing slippage into the cutting since there would be no component of dip at right angles to the face of the cutting. Factors other than geological ones determine the siting and direction of cuttings. Frequently they are not parallel to the dip of the strata and, as a result, we see an apparent dip in the cutting sides.

In the event of a geological section being at right angles to the direction of dip it will be, of course, in the direction of strike of the beds. There will be no component of dip seen in this section and the beds will appear to be horizontal. (Of course, close examination of such a quarry face or cutting will reveal that the strata are not horizontal but are dipping towards or away from the observer.)

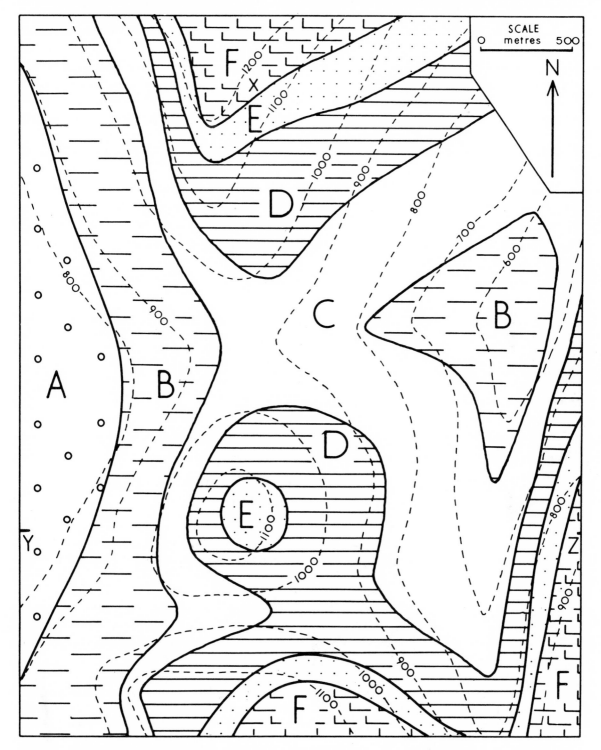

Map 4 Draw structure contours on the geological boundaries. Give the gradient of the beds (dip). Draw a section along the east–west line Y–Z. Calculate the thicknesses of beds B, C, D and E. Indicate on the map an inlier and an outlier.

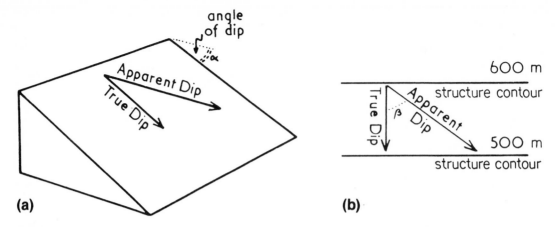

Fig. 6 (a) Diagram and (b) plan or map of structure contours to illustrate the relationship between true and apparent dip.

Calculation of the thickness of a bed

On Map 4 it can be seen that the 1100 m structure contour for the geological boundary D–E coincides with the 1000 m structure contour for boundary C–D. Thus, along this strike direction, the top of bed D is 100 metres higher than its base. It has a vertical thickness of 100 metres. This is the thickness of the bed that would be penetrated by a borehole drilled at point X.

Turn back to Map 3. The 200 metre structure contour for the Q–R boundary, which has already been inserted on the map, passes through the

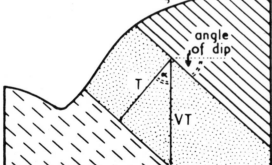

Fig. 7 Section showing the relationship between the vertical thickness (VT) and the true thickness (T) of a dipping bed.

point where the P–Q boundary is at 400 metres. Since structure contours are all parallel – where beds are simply dipping as in this map – this line is also the 400 metre structure contour for the P–Q boundary. It follows that bed Q has a vertical thickness of 200 metres. You will find that on many problem maps bed thickness is often 100 metres, 200 metres or some multiple of 50 metres in order to produce a simpler problem.

Vertical thickness and true thickness

Since the beds are inclined, the vertical thickness penetrated by a borehole is greater than the true thickness measured perpendicular to the geological boundaries (interfaces) (Fig. 7). The angle α between VT (vertical thickness) and T (true thickness) is equal to the angle of dip.

$$\text{Now Cosine } \alpha = \frac{T}{VT}$$
$$\therefore \mathbf{T} = VT \times \mathbf{Cosine}\ \alpha$$

The true thickness of a bed is equal to the vertical thickness multiplied by the cosine of the angle of dip. Where the dip is low (less than 5°) the cosine is high (over 0.99) and true and vertical thicknesses are approximately the same (see Table I, p. 66).

Width of outcrop

If the ground surface is level the width of outcrop of a bed of constant thickness is a measure of the dip (Fig. 8a).

Naturally, where beds with the same dip crop out on ground of identical slope the width of outcrop is related directly to the thickness of the beds (Fig. 8c).

More generally, beds crop out (or outcrop) on sloping ground and width of outcrop is a function of dip and slope of the ground as well as bed thickness. In Fig. 8b beds Y and Z have the same thickness (and dip) but, due to the different angles of intersection with the slope of the ground their widths of outcrop are very different. ($T_1 = T_2$ but W_1 is much greater than W_2.)

It will be noted that in the case of horizontal strata the geological boundaries are parallel to the topographic contours. In dipping strata the geological boundaries cross the topographic contours, and with irregular topography the steeper the dip the straighter the outcrops. In the limiting case, that of vertical strata, outcrops are straight and unrelated to the topography.

Inliers and outliers

An outcrop of a bed entirely surrounded by outcrops of younger beds is called an inlier. An outcrop of a bed entirely surrounded by older beds (and so separated from the main outcrop) is called an outlier. In Map 4 these features are the product of erosion on structurally simple strata and are called 'erosional' inliers and outliers. (See p. 35 for details of other inliers and outliers.)

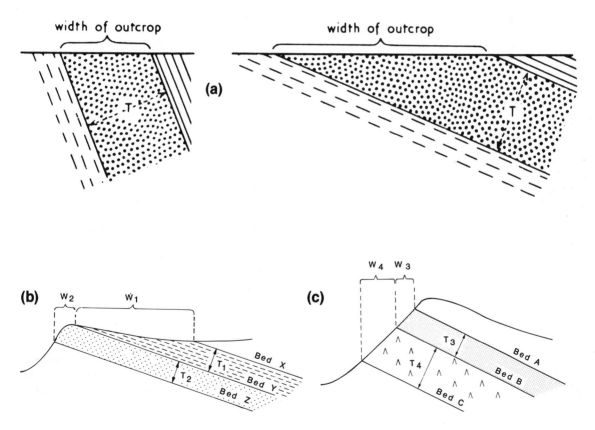

Fig. 8 (a) Sections showing the different widths of outcrop produced by a bed of the same thickness (T) with high dip and low dip. (b) Beds of the same thickness ($T_1 = T_2$) outcropping on differing slope. (c) Beds of different thickness ($T_4 = T_3 \times 2$) outcropping on a uniform slope.

Vertical Exaggeration It is very difficult to make the vertical scale equal to the horizontal scale for a one-inch to the mile map, which is of course, $1'' = 5,280$ feet. It is necessary to introduce a vertical exaggeration which should be kept to the minimum practicable. No hard and fast rule can be made. The British Geological Survey commonly employ a Vertical Exaggeration of two or three, but sections provided on BGS maps range from true scale to a vertical exaggeration of as much as x10 where this is necessary. The immensely useful 'Geological Highway Map' series of the USA (published by and obtainable from the AAPG, Tulsa, Oklahoma) illustrate the effects of an overlarge VE of $\times 20$. With one inch to the mile maps a vertical scale of $1'' = 1,000$ feet is particularly convenient and the VE at approximately $\times 5\frac{1}{4}$ is generally acceptable.

It is essential to realize that where the section is drawn with a vertical exaggeration of three, for example, the *tangent* of the angle of dip must be multiplied by three to find the angle of dip appropriate to this section. (It is *not the angle of dip* which is multiplied by three.) A dip of 20° should be shown on a section with a VE of $\times 3$ as a dip of 48°. Look up the tangents and check that this is correct.

Sections across published Geological Survey Maps
Henley-on-Thames: 1:50,000 (sheet 254) Solid & Drift Edition Examine the map and section, noting the relationship of topography to geology. (The Chalk is one of the most resistant formations in South East England forming the high ground of the Chilterns, Downs, etc.) Note numerous outliers to the west of the main cuesta.

A structure contour map is included. You will see that the structure contours are approximately parallel but not straight as on most problem maps which cover a smaller area as rule. What do you conclude from the curvature of the structure contours?

Aylesbury: 1'' (Sheet 238) Excluding the rather extensive Pleistocene and Recent deposits (which form a quite thin superficial cover), the oldest strata are to be found in the north-west of the area with successively younger beds to the south-east. This gives the direction of dip. The amount of dip can best be determined by ensuring that beds are made the appropriate thickness on the section, e.g. Lower Chalk and Middle Chalk should each measure approximately 200 feet if the section has been correctly drawn. Draw a section along a line in a north-west–south-east direction across the map to illustrate the structure of the area.

Note on the Aylesbury and Henley-on-Thames sheets. When dealing with an area the size of one of these maps it is found that strata are not uniformly dipping (as is the case on the simpler problem maps) but may be slightly flexured. Structure contours – if they could be drawn – would not be quite straight nor precisely parallel. The beds cannot be inserted on a section by constructing structure contours: they must be 'fitted' to the outcrop widths and drawn at their correct thicknesses, using information given in the stratigraphic column in the margin of the map.

Notes on BGS Maps The British Geological Survey produces a much wider variety of maps than formerly, both in range of scales employed and in the kind of information included.

Most of England and parts of Wales, together with about half of Scotland has now been covered by maps published on the 1:50,000 scale. Some areas are covered by still available and most useful one-inch to the mile sheets. Both solid geology editions and drift editions have been produced for some areas. For the purposes of investigating geological structures the solid editions are the more useful. However, drift editions, showing the superficial deposits, are of vital importance to the engineering geologist concerned with planning motorways, dams and foundations. In areas where drift deposits are not extensive a single edition combining Solid and Drift is usually published.

2 'Three-point' problems

If the height of a bed is known at three or more points (not in a straight line), it is possible to find the direction of strike and to calculate the dip of the bed, provided dip is uniform. This principle has many applications to mining, opencast and borehole problems encountered by applied geologists and engineers but this chapter deals only with the fundamental principle and includes a few simple problem maps.

The height of a bed may be known at points where it outcrops or its height may be calculated from its known depth in boreholes or mine shafts. If the height is known at three points (or more) only one possible solution exists as to the direction and amount of dip, and this can be simply calculated.

Observe the height of the seam at points A, B and C where it outcrops. Join with a straight line the highest point on the coal seam, C (600 m) to the lowest point on the seam, A (200 m). Divide the line A–C into four equal parts (since 600 m – 200 m = 400 m). As the slope of the seam is constant we can find a point on AC where the seam is at a height of 400 m (the mid-point). We also know that the seam is at a height of 400 m at point B. A straight line drawn through these two points is the 400 m structure contour. On a simply dipping stratum such as this, all structure contours are parallel. Construct the 200 m structure contour through point A, the 300 m, the 500 m and the 600 m structure contour – the latter through point C. Having now established both the direction and the spacing of the structure contours, complete the pattern over the whole of the map.

Construction of structure contours

Note: since Map 5 portrays a coal seam, and an average seam is of the order of 2 metres or less in thickness, on the scale of 2.5 cm = 500 m its thickness is such that it can satisfactorily be represented by a single line on the map. There is no need to attempt to draw structure contours for the top and the base of the seam – on this scale they are essentially identical.

Depth in boreholes

Relative to sea level, the height of the ground at the site of a borehole can be estimated from its proximity to contour lines and that of the coal seam at the same point on the map can be calculated from the structure contours. Quite simply, the difference in height between the ground surface and the seam is the depth to which the borehole must be drilled to reach the seam.

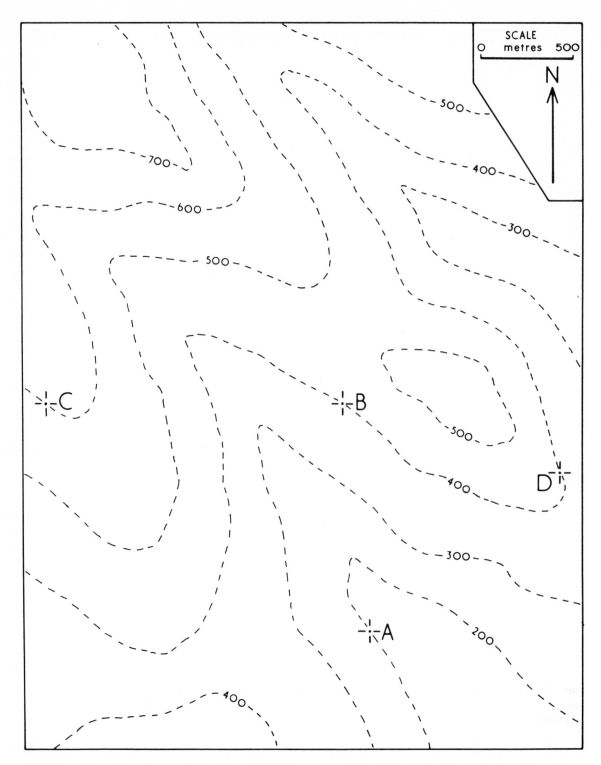

Map 5 Deduce the dip and strike of the coal seam which is seen to outcrop at points A, B and C. At what depth would the seam be encountered in a borehole sunk at point D? Complete the outcrops of the seam. Would a seam 200 m below this one also outcrop within the area of the map? Contours in metres.

Insertion of outcrops

The structure contours were drawn by ascertaining the height of the coal seam where it outcropped on contour lines. Wherever the seam – defined by its structure contours – is at the same height as the ground surface – defined by topographic contour lines – it will outcrop. We can find on the map a number of intersections at which structure contours and topographic contours are of the same height: the outcrop of the seam must pass through all these points.

Further, these points cannot be joined by straight lines. We must bear in mind that where the seam lies between two structure contours, e.g. the 300 m and 400 m, it can only outcrop where the ground is also at a height of between 300 m and 400 m, i.e. between the 300 m and 400 m topographic contours (Fig. 9). The outcrop of a geological boundary surface cannot, on a map, cross a structure contour or a topographic contour line except where they intersect at the same height.

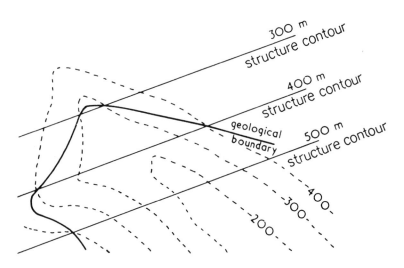

Fig. 9 The insertion of a geological boundary on a map with topographic contours and structure contours.

Map 6 Borehole A passes through a coal seam at a depth of 50 m and reaches a lower seam at a depth of 450 m. Boreholes B and C reach the lower seam at depths of 150 m and 250 m respectively. Having determined the dip and strike, map in the outcrops of the two seams (assume that the seams have a constant vertical separation of 400 m). Indicate the areas where the upper seam is at a depth of less than 50 m below the ground surface. It is necessary to first calculate the height (relative to sea level) of the lower coal seam at each of the points A, B and C where boreholes are sited.

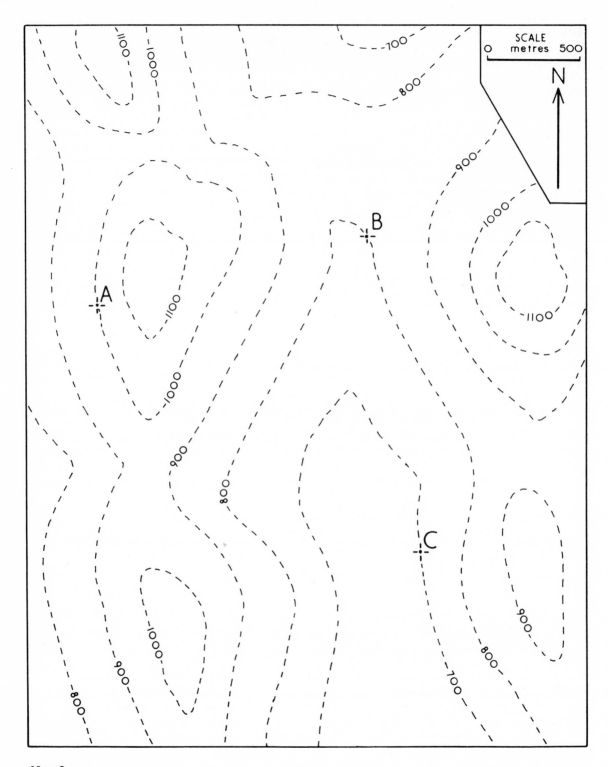

Map 6

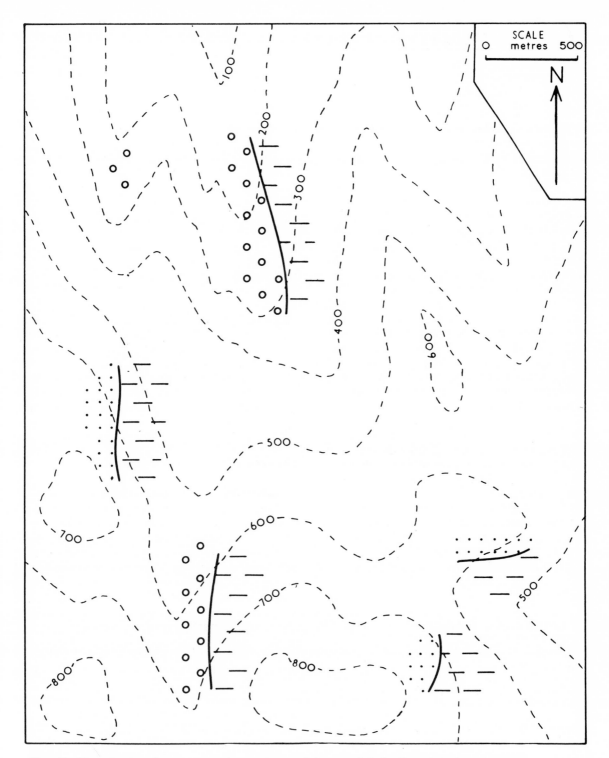

Map 7 Three beds outcrop – conglomerate, sandstone and shale. Complete the geological boundaries between these beds, assuming that the beds all have the same dip. Indicate on the map an inlier and an outlier.

3 Unconformities

In terms of geological history an unconformity represents a period of time during which strata are not laid down. During this period, strata already formed may be uplifted and tilted by earth-movements which also terminated sedimentation. The uplifted strata, coming under the effects of sub-aerial weathering and erosion, are 'worn down' to a greater or lesser extent before subsidence causes the renewal of sedimentation and the formation of further strata. As a result we find, in the field, one set of strata resting on the eroded surface of an older set of beds.

Many of the features which characterize unconformities cannot be deduced from map evidence alone. For example, only in the field can one observe the actual erosion surface, the presence of palaeosols, hardgrounds, etc. Derived fragments of the older strata may be redeposited in the post-unconformity strata (sometimes as a basal conglomerate) and the analogous but much rarer phenomenon of derived fossils may be seen. However, the evidence of a gap in the stratigraphic succession should be indicated in the stratigraphic column provided on a map and some of the above features may be referred to in the data given.

On a map the main evidence of unconformity is a difference of dip and strike direction in the pre- and post-unconformity strata. Earth movements which, by uplift, terminated sedimentation and subsequent sinking which permitted resumed sedimentation frequently resulted in differences of dip and strike.

An exception to this can be found on the margins of the London Basin, where the Lower Tertiary beds rest unconformably on the Chalk with little difference in strike or dip yet, by comparison with successions of strata on the other side of the Channel, we know that this is a major unconformity representing a long period of time since, in Britain, the uppermost stage of the Chalk and the lowest two stages of the Tertiary are absent.

In some cases an unconformity represents a time interval of such length that the older strata were intruded by igneous rocks or were subjected to metamorphism.

Overstep

Usually the lowest bed of the younger series of strata, having a quite different dip and strike from that of the older strata, rests on beds of different age. This feature (Fig. 10) is called overstep; bed X is said to overstep beds A, B, C, etc.

If the older strata were tilted before erosion took place, they meet the plane of unconformity at an angle, and there is said to be an 'angular unconformity' (Fig. 10).

Overlap

As subsidence continues and the sea for example spreads further on to the old land area, successive beds are laid down, and they may be of greater geographical extent, so that a particular bed spreads beyond, or overlaps, the preceding bed. This feature (Fig. 11) may accompany an unconformity with or without overstep. Bed Y overlaps Bed X. (The converse effect, that of successive beds being laid down over a progressively contracting area of deposition, due to gradual uplift, is known as off-lap. Such a feature is rarely deducible from a geological map and will be discussed no further.)

Map 8 Find the plane of unconformity. Deduce the direction and amount of dip of the two series of beds. Draw a section along the line from the north-west corner of the map to the south-east corner. Would the coal seam be encountered in boreholes situated at points A, B and C? If the coal is present calculate its depth below the ground surface, if it is absent suggest an explanation for its absence. Indicate the position of the coal seam beneath bed Y.

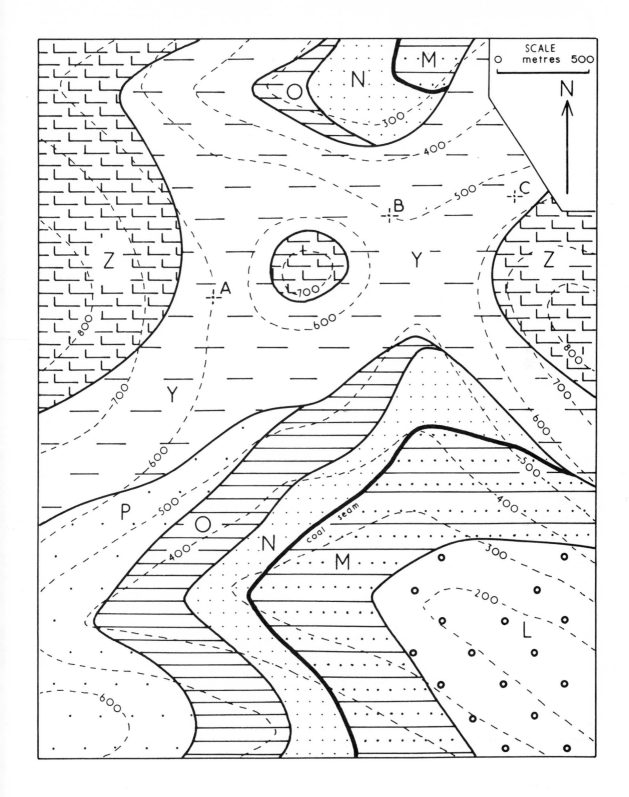

Map 8

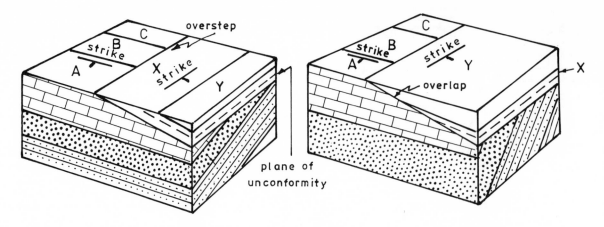

Fig. 10 Block diagram of an angular unconformity. **Fig. 11** Block diagram of an unconformity with overlap.

Two principal types of unconformity are recognizable:

(1) planes of marine erosion, readily definable by structure contours.

(2) buried landscapes. Sub-aerial erosion had not produced a peneplain and the post-unconformity sediments were deposited on a very irregular surface of hills and valleys, gradually burying these features.

Problem maps 8, 16 and 20 are exercises based on type (1). Usually, it is necessary to look at a broader regional geology to see the effects of buried landscape. See below the note on the Assynt one-inch geological sheet.

Sub-unconformity outcrops

An unconformity represents a period of erosion. Tens or hundreds of metres of strata may be removed. How can we examine this phenomenon and deduce what remains of the older strata –

since they are now covered by post-unconformity strata? We could, of course, in practice remove all post-unconformity strata with a bulldozer and other earth-moving equipment to reveal the plane of unconformity. We can, by a study of the structure contours, deduce from a map just what we should expect to see if that was possible.

What we seek are the 'outcrops' of the older strata on the plane of unconformity. This plane can be defined by its structure contours. If we now take the structure contours drawn on the geological boundaries of the older set of strata and note where they intersect, not ground contours, but the structure contours of the plane of unconformity, we can plot – where the two sets of structure contours intersect at the same height – a number of points which will define the sub-unconformity outcrops. This exercise can be attempted on Map 8. The topic will be revised on a later page and Maps 10 and 16 give further practical exercises.

Sections across published Geological Survey Maps
Assynt 1″ Geological Survey special sheet Examine the western part of the map to find the unconformities at the base of the Torridonian and the base of the Cambrian. Draw a section along the north–south grid line 22 to show these unconformities. From your knowledge of the conditions under which the Torridonian and Lower Cambrian were deposited can you explain how and why these unconformities differ?

Shrewsbury 1″ Map No. 152 Examine the map carefully. How many of the unconformities indicated in the geological column in the margins are deducible from the map evidence?

18

4 Folding

We have seen that strata are frequently inclined (or dipping). On examining the strata over a wider area it is found that the inclination is not constant and, as a rule, the inclined strata are part of a much greater structure. For example, the Chalk of the South Downs dips generally southwards towards the Channel – as can be determined by examination of the 1:50,000 Geological Survey map of Brighton (Sheet No. 315). We know, however, that in the North Downs the Chalk dips to the north (passing beneath the London Basin); the inclined strata of the South and North Downs are really parts of a great structure which arched up the rocks including the Chalk over the Wealden area. Not all arching of the strata is of this large scale and minor folding of the strata can be seen to occur near the centre of the Brighton sheet.

Folding of strata represents a shortening of the earth's crust and results from compressive forces.

Anticlines and synclines

Where the beds are bent upwards into an arch the structure is called an anticline (*anti* = opposite: *clino* = slope; the beds dip away from each other

on opposite sides of the arch-like structure). Where the beds are bowed downwards the structure is called a syncline (*syn* = together: *clino* = slope; beds dip inwards towards each other) (Fig. 12). In the simplest case the beds on each side of a fold structure, i.e. the limbs of the fold, have the same amount of dip and the fold is symmetrical. In this case a plane bisecting the fold, called the axial plane, is vertical. The fold is called an upright fold whenever the axial plane is

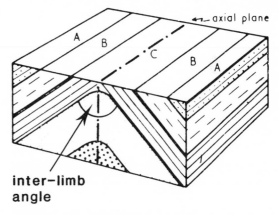

Fig. 13 Block diagram of a symmetrical anticline (an upright fold).

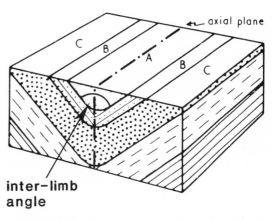

Fig. 14 Block diagram of a symmetrical syncline (an upright fold).

Fig. 12 Diagrammatic section of folded strata.

vertical or steeply dipping. (Where axial planes have a low dip or are nearly horizontal, see Fig. 38a, folds are called flat folds.)

The effect of erosion on folded strata is to produce outcrops such that the succession of beds of one limb is repeated, though of course in the reverse order, in the other limb. In an eroded anticline the oldest bed outcrops in the centre of the structure and, as we move outwards, successively younger beds are found to outcrop (Fig. 13). In an eroded syncline, conversely, the youngest bed outcrops at the centre of the structure with successively older beds outcropping to either side (Fig. 14).

Asymmetrical folds

In many cases the stresses in the earth's crust producing folding are such that the folds are not symmetrical like those described above. If the beds of one limb of a fold dip more steeply than the beds of the other limb, then the fold is asymmetrical. The differences in dip of the beds of the two limbs will be reflected in the widths of their outcrops, which will be narrower in the case of the limb with the steeper dip (Fig. 15). (See also p. 9 and Fig. 8). Now, the axial plane bisecting the fold is no longer vertical but is inclined and the fold is called an inclined fold.

Overfolds If the asymmetry of a fold is so great that both limbs dip in the same direction (though with different angles of dip), that is to say the steeply dipping limb of an asymmetrical fold has

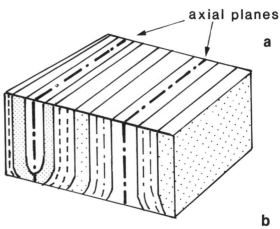

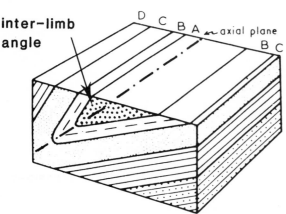

Fig. 15 Block diagram of an asymmetrical syncline (an inclined fold).

Fig. 16 Block diagram of an overfold (an overfolded syncline).

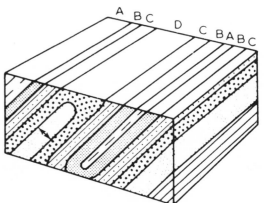

Fig. 17 Block diagrams of isoclinal folding: (a) upright folds (b) overfolds

20

been pushed beyond the vertical so that it has a reversed – usually steep – dip, the fold is called an overfold (Fig. 16). The strata of the limb with the reversed dip, it should be noted, are upside down, i.e. inverted.

Figures 14 to 17 show a progressive decrease in inter-limb angle.

Figures 14 to 17 illustrate fold structures produced in response to increasing tectonic stress. Further terminology should be noted. Where the limbs of a fold dip at only a few degrees it is a gentle fold, with somewhat greater dip (Figs. 13, 14) a fold is described as open, with steeper dipping limbs a fold is described as close and with parallel limbs (Fig. 17a, b) the folding is isoclinal.

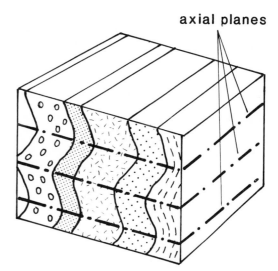

axial planes

Fig. 18 In an area of more complex structural geology, not produced simply by one epoch of compression, flat folds such as these may be found. They are open folds with horizontal or near-horizontal axial planes.

Isoclinal folds Isoclinal folds are a special case of over-folding in which the limbs of a fold both dip in the same direction at the same angle (*isos* = equal: *clino* = slope), as the term suggests (Fig. 17b). The axial planes of a series of such folds will also be approximately parallel over a small area, but over a larger area extending perhaps forty kilometres (greater than that portrayed in a problem map) they may be seen to form a fan structure.

Similar and concentric folding

When strata, originally horizontal, are folded it is clear that the higher beds of an anticline form a greater arc than the lower beds (and the converse applies in a syncline). Theoretically at least two mechanisms are possible: the beds on the outside of a fold may be relatively stretched while those on the inside are compressed, or the beds on the outside of a fold may slide over the surface of the inner beds (Fig. 19).

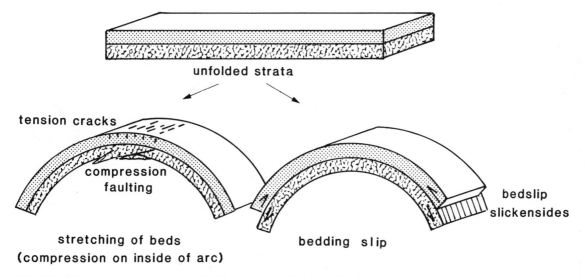

Fig. 19 The response of strata to folding, two possible mechanisms.

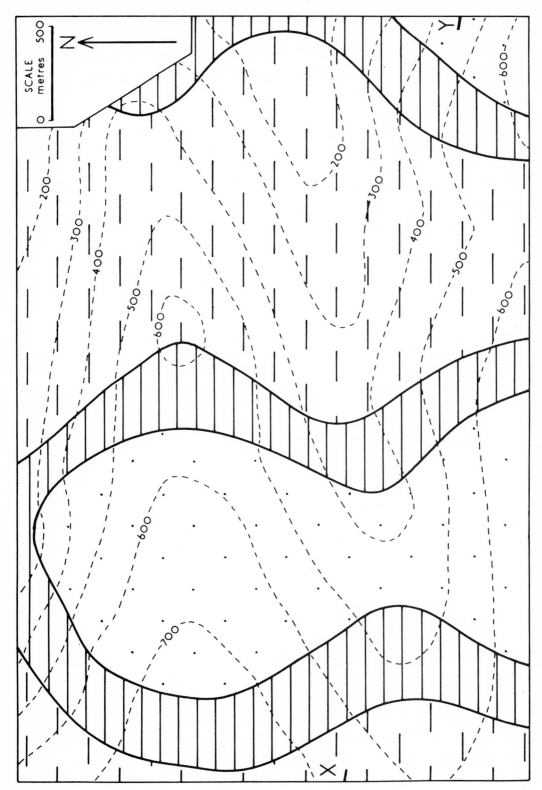

Map 9 Draw structure contours for the upper and lower surface of the shaded bed of shale. Is the direction of strike approximately north–south or east–west? Indicate on the map the position of an anticlinal axial trace and the position of a synclinal axial trace. Draw a section along the line X–Y. (The axial trace is the outcrop of the axial plane.)

22

The way in which beds will react to stress depends upon their constituent materials (and the level in the crust at which the rocks lie). Competent rocks such as limestone and sandstone do not readily extend under tension or compress under compressive forces but give way by fracturing and buckling while incompetent rocks such as shale or clay can be stretched or squeezed. Thus in an alternating sequence of sandstones and shales the sandstones will fracture and buckle while the shales will squeeze into the available spaces.

Concentric folds   The beds of each fold are approximately concentric, i.e. successive beds are bent into arcs having the same centre of curvature. Beds retain their constituent thickness round the curves and there is little thinning or attenuation of beds in the limbs of the folds (Fig. 20a).

Straight limbed folds also maintain the uniformity of thickness of the beds (except in the hinge of the fold) and folding takes place by slip along the bedding planes as it does in the case of concentric folds. Although typically developed in thinly bedded rocks (such as the Culm Measures of North Devon) most of the problem maps in this book which illustrate folding have straight limbed folds since these provide a simple pattern of equally spaced structure contours on each limb of a fold.

Similar folds   The shape of successive bedding planes is essentially similar, hence the name (Fig. 20b). Thinning of the beds takes place in the limbs of the folds (and a strong axial plane cleavage is usually developed). This type of folding probably occurs when temperatures and pressures are high.

(a)

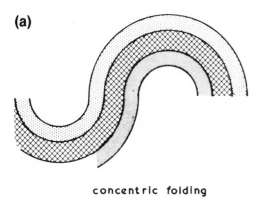

concentric folding

(b)

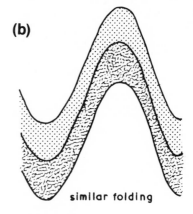

similar folding

Fig. 20   The shape of concentric and similar folding seen in section.

Two possible directions of strike

A structure contour is drawn by joining points at which a geological boundary surface (or bedding plane) is at the same height. By definition this surface is at the same height along the whole length of that structure contour. Clearly, if we join points X and Y (Fig. 21) we are constructing a structure contour for the bedding plane shown, for not only are points X and Y at the same height but the bedding plane is at the same height along

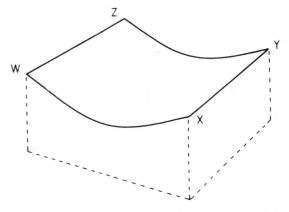

Fig. 21   Block diagram to illustrate the direction of structure contours in folded strata.

23

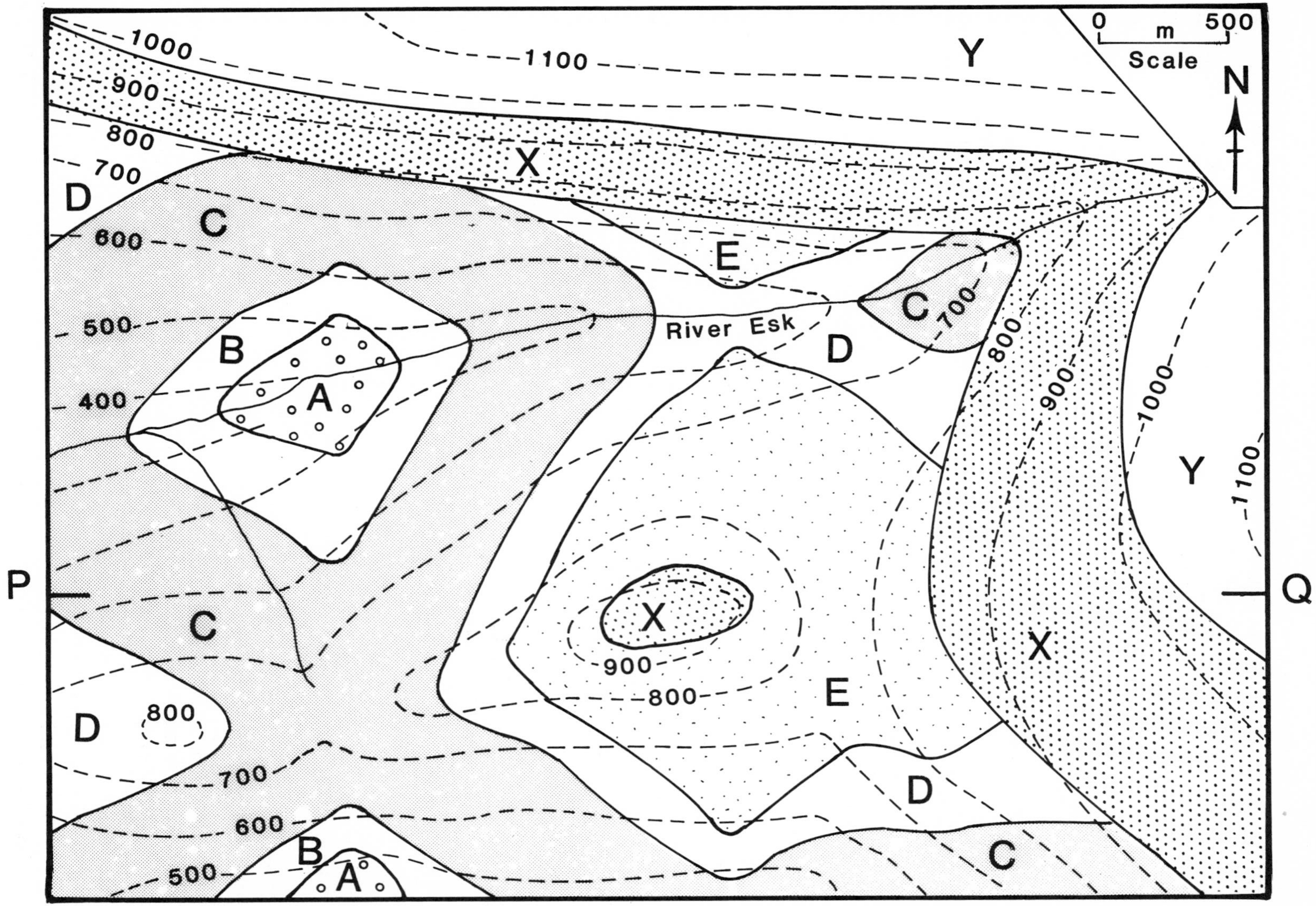

N

Scale

500 m 0

1100

1000

900

800

700

Y

X

River Esk

C

D

E

C

D

E

X

900

800

B

A

B

A

C

D

800

600

500

400

D

C

700

600

500

X

E

D

C

Y

1100

1000

900

800

700

1000

900

800

700

P

Q

Map 10

24

the line X–Y. If, however, we join the points W and X, although they are at the same height, we are not constructing a structure contour, for the bedding plane is not at the same height along the line W–X; it is folded downwards into a syncline. Thus, if we attempt to draw a structure contour pattern which proves to be incorrect, we should look for the correct direction approximately at right angles to our first attempt. It should also be noted that an attempt to visualize the structures must be made. For example, in Map 9 the valley sides provide, in essence, a section which suggests the synclinal structure, especially if the map is turned upside down and viewed from the north. Map 11 similarly reveals the essential nature of the structures by regarding the northern valley side as a section. To facilitate this, fold the map at right angles along the line of the valley bottom. Now regard the top half of the map as an approximate geological section.

What is the test of whether we have found the correct direction of strike? In these relatively simple maps the structure contours should be parallel and equally spaced (at least for each limb of a fold structure). Furthermore, calculations of true thicknesses of a bed at different points on the map should give the same value.

Note on Map 10   You should find three areas where Bed E occurs underneath younger beds. The eastern extent of Bed E could be deduced by joining the three points where the D/E junction is cut by the base of Bed X. However, to confirm that this is correct it is necessary to use the intersection of the two sets of structure contours (on D/E and on the base of X). This method is the only way in which the western extent of Bed E can be defined.

Remember (see p. 18 'Sub-unconformity outcrops') that the topography is irrelevant to the solution of this part of the problem. The surface on which we are plotting the outcrop of the D/E boundary is the plane of unconformity, defined by the structure contours drawn on the base of Bed X. You will find it necessary to use intersections of structure contours which are now above ground level: although the strata in question have now been eroded away from these points, the points themselves remain valid as construction points.

Notes on the Lewes map   Here, as elsewhere in the south-east of England, the Chalk of Upper Cretaceous age is a compact rock more resistant to erosion than the softer clays and sands of pre-Chalk age. There is less information on dip of strata than desirable, although dips are given to the north and south of the town of Lewes. Remember to allow for the vertical exaggeration by multiplying the gradient (tangent) of given dips by 4. Where dip information is less than adequate the beds must be fitted to the outcrops at an angle of dip which gives the correct thicknesses of strata shown in the stratigraphic column on the map.

Published Geological Survey Maps
Lewes, 1:50,000, BGS Map (Sheet No. 319) Solid & Drift edition, 1979 Study the geology of this area on the southern limb of the Wealden anticline. Note the general structural trend, close to E–W, and the 'younging' of beds southwards. Study the relationship of topography to geology. Draw a section along the line of Section 1 (south of grid line 117). Make the vertical exaggeration x4 (the same as the BGS section on this map).

Map 10   Indicate on the map the outcrop of the plane of unconformity. Insert the axial traces of the folds. Draw a section along the line P–Q to illustrate the geology. Indicate on the map the extent of Bed E beneath the overlying strata.

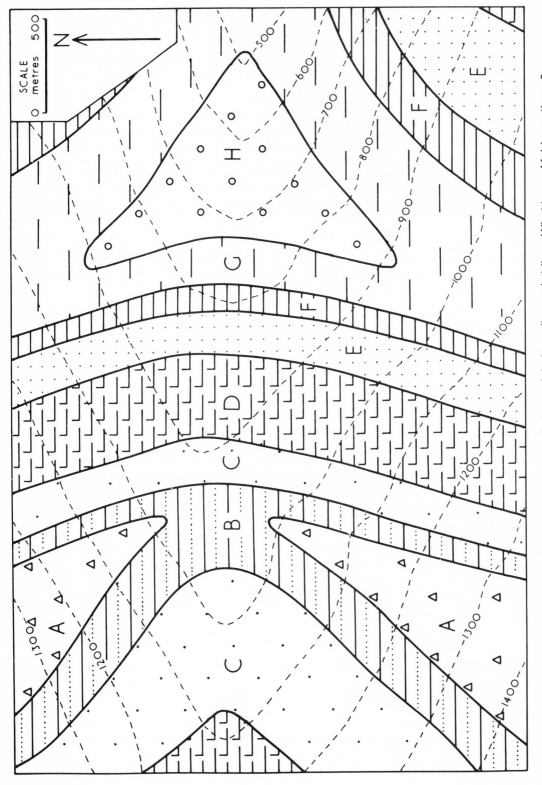

Map 11 Draw structure contours on all the geological boundaries and deduce dips and strikes. What type of folds are these? Note that all geological boundaries 'V' downstream. Draw a section along an east–west line. Draw in the axial traces, i.e. the outcrops of the axial planes of the folds.

5 Faults

The previous chapter showed how geological strata subjected to stress may be bent into different types of folds. Strata may also respond to stress (compression, extension or shearing) by fracturing.

Faults are fractures which displace the rocks. The strata on one side of a fault may be vertically displaced tens, or even hundreds, of metres relative to the strata on the other side. In another type of fault the rocks may have been displaced horizontally for a distance of many kilometres. While in nature a fault may consist of a plane surface along which slipping has taken place, it may on the other hand be represented by a zone of brecciated (composed of angular fragments) rock. For the purposes of mapping problems it can be treated as a plane surface, usually making an angle with the vertical.

All structural measurements are made with reference to the horizontal, including the dip of a fault plane. This is a measure of its slope, cf. the dip of bedding planes, etc. The term 'hade', formerly used, is the angle between the fault plane and the vertical (and is therefore the complement of the dip). It causes some confusion but it will be widely encountered in books and on maps and is,

for this reason, shown in Figures 22 and 23. It will, no doubt, gradually drop out of usage.

The most common displacement of the strata on either side of a fault is in a vertical sense. The vertical displacement of any bedding plane is called the throw of the fault. Other directions of displacement are dealt with later in this chapter.

Normal and reversed faults

If the fault plane is vertical or dips towards the downthrow side of a fault it is called a normal fault (Fig. 22). If the fault plane dips in the opposite direction to the downthrow (i.e. towards the upthrow side) it is a reversed fault (Fig. 23).

In nature, the dip of a reversed fault is generally lower than that of a normal fault. It may be less than 45°. In an area of strong relief the outcrop of a reversed fault may be sinuous. The outcrop of a normal fault (commonly with a dip in the 85–75° range, but as low as 50° in some examples) will usually be much straighter – unless the fault-plane itself is curved (see Map 13). Of course, where the area is of low relief, the outcrop of a fault plane, normal or reversed, will be virtually straight.

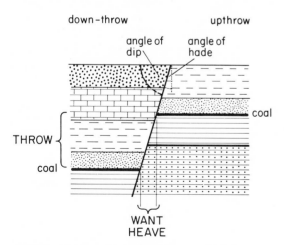

Fig. 22 Section through strata displaced by a normal fault (after erosion has produced a near-level ground surface).

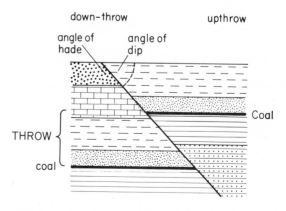

Fig. 23 Section through strata displaced by a reversed fault.

On some geological maps, such as those produced in Canada and the United States, the direction of dip of fault planes is shown. The angle of dip of the fault plane may be given in the map description. British Geological Survey maps show the direction of the throw of faults by means of a tick on the downthrow side of the fault outcrop (see also Map 13).

Any sloping plane, including a fault plane, can be defined by its structure contours. It is possible on both Map 12 and 14 to construct contours for the fault plane. The method is exactly the same as for constructing structure contours on bedding planes, described in Chapter 1. From these structure contours the direction of dip of the fault plane can be deduced and then, by reference to the direction of downthrow, it can be deduced whether the fault is normal or reversed.

The effects of faulting on outcrops

Consider the effects of faulting on the strata: those on one side of a fault are uplifted, relatively, many metres. Since this uplift is not as a rule a rapid process and the strata will be eroded away continuously, a fault may not make a topographic feature, although temporarily a fault scarp may be present (Fig. 24) especially after sudden uplift resulting from an earthquake.

Some faults which bring resistant rocks on the one side into juxtaposition with easily eroded rocks on the other side may be recognized by the presence of a fault line scarp (cf. fault scarp resulting from the actual movement).

The strata which have been elevated on the upthrow side of a fault naturally tend to be eroded more rapidly than those on the downthrow side. This results in higher (younger) beds in the strati-graphical sequence being removed from the upthrow side of the fault while they are preserved on the downthrow side. It follows that we can usually determine the direction of downthrow of a fault, whether normal or reversed. Following the line of a fault across a map, there will be points where a younger bed on one side of the fault is juxtaposed against an older bed on the other side of the fault. The younger bed will be on the downthrow side of the fault.

A fault dislocates and displaces the strata. The effect of this, in combination with erosion, is to cause discontinuity or displacement in the outcrops of the strata.

Classification of faults

Faults may be categorized in two ways.
1 Faults may be classified according to the direction of displacement of the blocks of strata on either side of the fault plane. We have considered so far normal and reversed faults with a vertical displacement called throw. Movement in these faults was in the direction of dip of the fault plane. They are called dip-slip faults because the movement – or displacement – was parallel to the direction of dip of the fault plane.

On a later page faults with lateral displacement, wrench faults, are described. Here, displacement is parallel to the strike of the fault plane and they can be described as strike-slip faults.

In nature, in some faults the displacement was neither dip-slip nor strike-slip but was oblique. Naturally, the displacement will have a vertical component (throw) and a horizontal component (lateral displacement). Such a fault may be called an oblique-slip fault (Fig. 25).

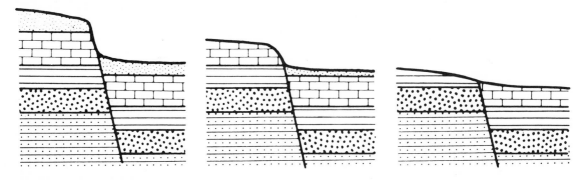

Fig. 24 Sections to show the progressive elimination of a fault scarp by erosion.

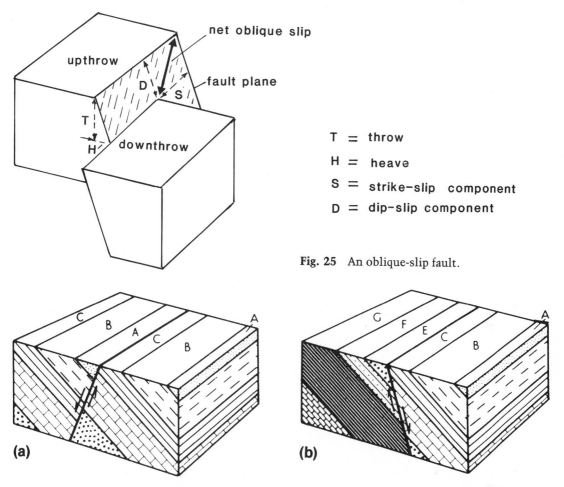

net oblique slip

upthrow

fault plane

D

S

T

H

downthrow

T = throw

H = heave

S = strike-slip component

D = dip-slip component

Fig. 25 An oblique-slip fault.

(a)

(b)

Fig. 26 Block diagrams of a normal strike fault (a) with the direction of bedding dip opposite to the direction of the dip of the fault plane, causing repetition of part of the succession of outcrops and (b) with the dips of bedding and the fault plane in the same direction, causing a suppression of part of the succession of outcrops.

2 A different classification of faults is dependent upon their geographical pattern, especially in relation to the dip directions of the strata cut by the faults. Where the faulting is parallel or nearly so to the direction of dip of strata, the faults are called dip faults. Where the faulting is more or less at right angles to the direction of dip of strata, i.e. approximately parallel to their strike, the faults are called strike faults. Examples of both are found on Problem Map 13. Faults which are neither in the dip direction nor the strike direction may be called oblique faults.

It is particularly important not to confuse the two schemes of classification discussed above: the terms are unfortunately very similar. Test yourself: what is a dip-slip strike fault? Figures 26 and 27 show examples of dip-slip faults, the former shows two cases of strike faults, the latter shows a dip fault. Figure 29 is an example of a strike-slip fault. It is, however, also a dip fault.

To return to normal dip-slip faults, sequences of outcrops encountered on a traverse may be partly repeated or may be partly suppressed.

Where the fault plane is parallel to the strike of the beds we see either repetition of outcrops (Fig. 26a) where the succession of beds at the surface reads A, B, C, A, B, C or the suppression of outcrops (Fig. 26b) where the succession of beds at the surface reads A, B, C, E, F, G.

Where the fault plane is parallel to the dip direction of the strata (a dip fault), i.e. at right angles to the strike, a lateral shift of the outcrop occurs.

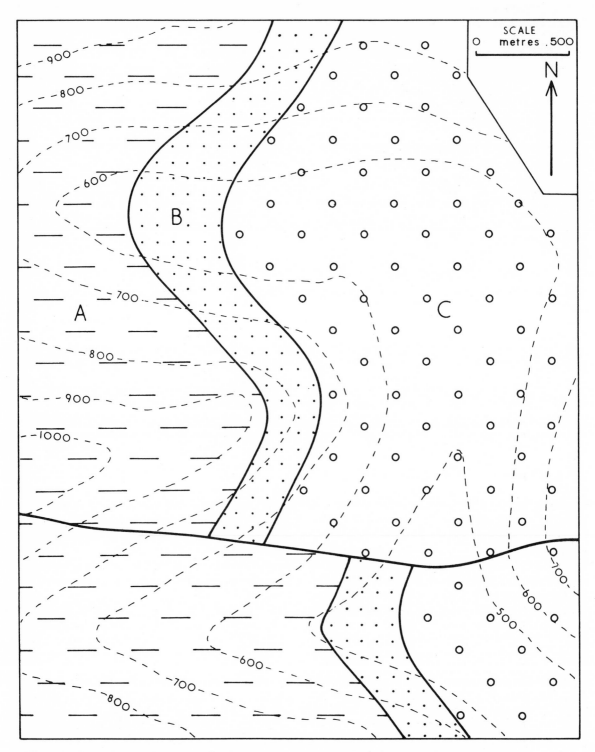

Map 12 Draw structure contours for the upper and lower surfaces of the sandstone (stippled). What is the amount of the throw of the fault? Draw structure contours on the fault plane. Is it a normal or a reversed fault? What is the thickness of the sandstone?

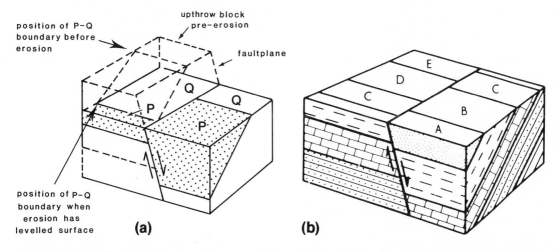

position of P–Q boundary before erosion

upthrow block pre-erosion

faultplane

Q

Q

P

P

position of P–Q boundary when erosion has levelled surface

(a)

E

D

C

C

B

A

(b)

Fig. 27 (a) Dip fault showing how vertical throw gives rise to lateral shift of outcrop as a result of erosion. The outcrops shift progressively down dip as erosion lowers the ground surface. (b) Block diagram of a normal dip fault. Note the lateral shift of outcrops although the actual displacement is vertical.

This must not be confused with lateral movement of the strata (see p. 33): the transposition of the outcrops is due to vertical displacement of the beds followed or accompanied by erosion which, because the strata are inclined, causes the outcrops on the upthrow side to be shifted in the direction of dip (Fig. 27).

Calculation of the throw of a fault

Note on Map 12 Construct the structure contours on the upper surface of the sandstone bed in the northern part of Map 12. They run north–south and are spaced at 12.5 mm intervals. Follow the same procedure for the upper surface of the sandstone south of the fault plane. The 500 m structure contour drawn on the south side of the fault, if produced beyond the fault, is seen to be coincident with the position of the 1000 m structure contour on the north side of the fault. The stratum on the south side is therefore 500 m lower relatively. The fault has a downthrow to the south of 500 m.

Determine the throw of the fault using the structure contours on the base of the sandstone – on each side of the fault – and check that you obtain the same value.

Map 13 is a slightly simplified version of the western part of the BGS map of Chester, Sheet 109 in the 1:50,000 series, solid edition. It is here

about half scale. The geology of the area is relatively simple with strata dipping generally towards the east with dips in the range of 5 to 17 degrees. The topography is almost flat (and drift cover is extensive so that much of the map has been compiled using borehole data). Strata are displaced by a considerable number of faults and they display many important characters of faulting.

Note that faults may die out laterally, examples can be seen at F_1 and F_2. Of course, all faults die out eventually, unless cut off by another fault, as at C, though major faults may extend for tens or even hundreds of kilometres. Note also that a fault may curve, for example at G. This is not mere curvature of the outcrop of the fault plane due to the effects of topography on a sloping fault plane (see Map 14).

Most of the faults on Map 13 run approximately north–south, roughly parallel to the general strike of the strata: they can be called strike faults. The displacement is in the direction of dip of the fault planes so they can also be called dip-slip faults. Two faults are approximately parallel to the direction of dip of the strata, seen at C and D. These are, therefore, dip faults. They are also probably dip-slip faults. The Chester sheet indicates the direction of downthrow of each fault, as do most published maps, but it has been omitted from most of the faults on Map 13.

31

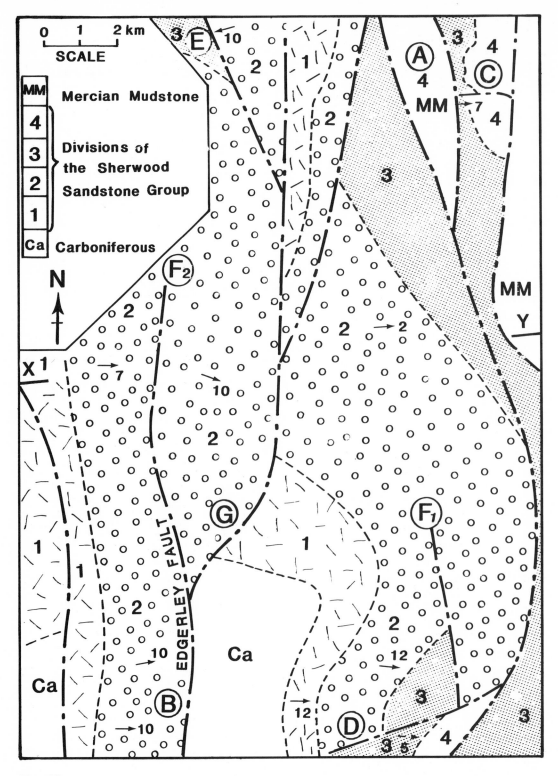

SCALE

0 1 2 km

MM Mercian Mudstone

4
3 Divisions of
2 the Sherwood
1 Sandstone Group

Ca Carboniferous

N

Map 13

32

It can be seen that since the fault plane dips, the intersections with a bedding plane on each side of the fault do not coincide (in plan view). Consider an economically important bed, for example of coal or ironstone. In the case of a normal fault there is a zone where a borehole would not penetrate this bed at all due to the effect of heave (see Fig. 22). This is important when calculating economic reserves, for example in a highly faulted coalfield. The estimate of reserves could be as much as 15% too great unless allowance had been made for fault heave. In the case of a reversed fault a zone exists where a borehole would penetrate the same bed twice. Figure 28 shows, in section, how a borehole after penetrating a coal seam would penetrate the fault plane and, beneath it, the same seam. In calculating reserves it is vital to recognize that it is the same seam which the borehole has encountered or reserve calculations would be wrong by a factor of approximately two.

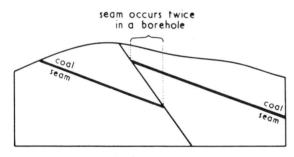

Fig. 28 Section to show a 'low angle' (high hade) reversed fault and its importance in mining problems.

Wrench or tear faults

In the case of these faults the strata on either side of the fault plane have been moved laterally relative to each other, i.e. movement has been a horizontal displacement parallel to the fault plane. In the case of simply dipping strata the outcrops are shifted laterally (Fig. 29) so that the effect, *on the outcrops*, is similar to that of a normal dip fault (cf.

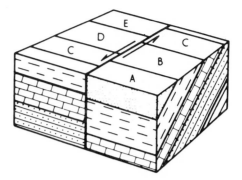

Fig. 29 Block diagram of a wrench (= tear) fault. Note that the effect on the outcrops is similar to that of a normal dip fault (cf. Fig. 27b).

Fig. 27) – and in this case it is usually impossible to demonstrate strike-slip from the map alone (apparent slip only can be found).

In some geological contexts the terminology based on the direction of movement relative to the dip and strike of the fault plane is preferable. Faults in which movement has been in the strike direction of the fault plane (= strike-slip faults) include wrench faults.

Notes on Map 13   The dip of the strata at E is anomalous, the result of a phenomenon called fault-drag.

The effect of faults which are parallel to the dip of the strata is to laterally shift outcrops (see p. 31 and Map 12). The extent of this shift depends on two factors, the amount of throw of the fault and the angle of dip of the strata. At C, on Map 13, the geological boundary is displaced very little, but at D the displacement is considerable (the shift is almost equal to the width of outcrop of Bed 4). At C and D the dips of the strata are similar so we can conclude that the fault at D has a much larger throw than the fault at C.

It may be assumed that all the faulting here is normal. Since we have no means of calculating the dip of the fault planes, in your section give faults a conventional dip towards the downthrow of 80 to 85°.

Map 13   On the map indicate the downthrow direction of all faults where it has not been shown. Indicate examples of (a) a graben and (b) step faulting. Draw a section along the nearly east–west line X–Y to illustrate the geology. Reproduced by permission of the Director General, British Geological Survey: NERC copyright reserved.

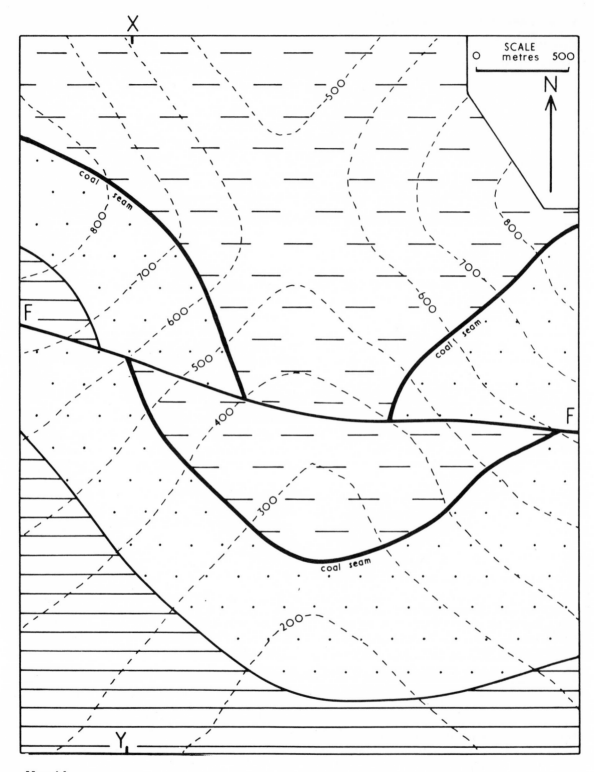

Map 14

Pre- and post-unconformity faulting

After the deposition of the older set of strata earth movements causing uplift may also give rise to faulting of the strata. The unconformable series (the younger set of beds), not being laid down until a later period, are unaffected by this faulting. Earth movements subsequent to the deposition of the unconformable beds would, if they caused faulting, produce faults which affect both sets of strata. Clearly, it is possible to determine the relative age of a fault from inspection of the geological map which will show whether the fault displaces only the older (pre-unconformity) strata or whether it displaces both sets of strata. A fault is later in age than the youngest beds it cuts.

A fault may also be dated relative to igneous intrusions, a topic dealt with in the last chapter.

Structural inliers and outliers

The increased complexity of outcrop patterns due to unconformity and faulting greatly increases the potential for the formation of outliers and inliers (these terms have been defined on p. 9). Indicate on Map 16 inliers and outliers which owe their existence to such structural features and subsequent erosional isolation.

Posthumous faulting

Further movement may take place along an existing fault plane. So the displacement of the strata is attributable to two or more geological periods. It follows that an older series of strata may be displaced by an early movement of the fault which did not affect newer rocks since they were laid down subsequently. The renewed movement along the fault will displace both strata so the older strata will be displaced by a greater amount since they have been displaced twice (the throws are added together).

Isopachytes

Isopachytes (*iso* = equal; *pakhus* = thick) are lines of equal thickness. The simplest use of isopachytes is to show the thickness of cover material overlying a bed of economic importance, such as a coal seam or ironstone. The overlying material – whatever its composition: strata, soil and subsoil – is called the overburden. Its thickness can be determined where the height of the top of an economic bed (ironstone, Map 15) is known from its structure contours, and the height of the ground at the same point is known from the topographic contours. Wherever structure contours and topographic contours intersect on the map we can obtain a figure for the thickness of overburden (by subtracting the height of the top of the ironstone from the height of the ground). Joining up the points of equal thickness gives an isopachyte. Where ironstone and ground are at the same height the thickness of overburden is nil and the bed must outcrop. (Its outcrop would be the 0 m isopachyte.) Bed (or stratum) isopachytes, concerned with beds of varying thickness, are dealt with on p. 45.

Note on Map 14   The zone in which a borehole penetrates the seam twice is defined by the lines of intersection of the fault plane and the coal seam (Fig. 28). The surfaces intersect where they are at the same height (where the coal seam structure contours and the fault plane structure contours of the same height coincide).

Map 14   The line F–F is the outcrop of a fault plane. The other thick line on the map is the outcrop of a coal seam. Shade areas where coal could be penetrated by a borehole (where it has not been removed by erosion). Indicate areas in which any borehole would penetrate the seam twice. What type of fault is this? Draw a section along the line X–Y. Draw the 100 m overburden isopachyte, i.e. a line joining all points where the coal is overlain by 100 m of strata.

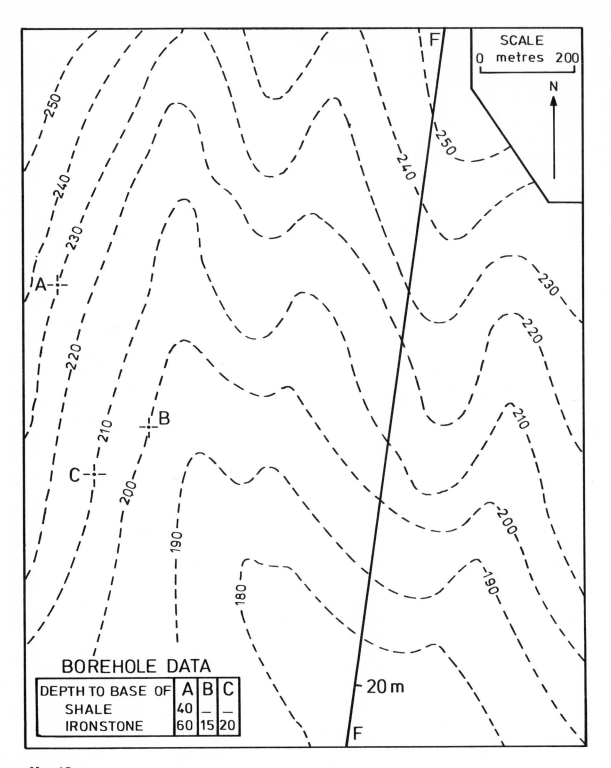

Map 15

Published Geological Survey Maps
Chesterfield: 1″ Geological Survey map (Sheet No. 112) Find the major unconformity on this map. Locate some faults which are older than the Permian beds and some which are younger. Draw a section along a line across the map ensuring that it passes through the Ashover anticline in the south-west.

Published Geological Survey Maps
Leeds 1:50 000 Map No. 70 Draw a section along the 'Line of Section' engraved on the map to show the geological structures.

Map 15 The western part of the map comprises a three-point problem enabling us to draw structure contours at 10 m intervals on the base of the ironstone. Assuming the top of the ironstone to be 20 m higher, why does borehole B penetrate only 15 m of ironstone? East of the fault (a normal fault of high angle of dip), produce the structure contours and re-number them 20 m lower (since the fault is shown as having a downthrow of 20 m to the east). Shade outcrops of iron ore on both sides of the fault. Also shade areas where the ironstone could be worked opencast if not more than 40 m of overlying shale is to be removed, i.e. draw the overburden isopachyte for 40 m. Draw a section along a north–south line passing through point B and another section along an east–west line passing through B.

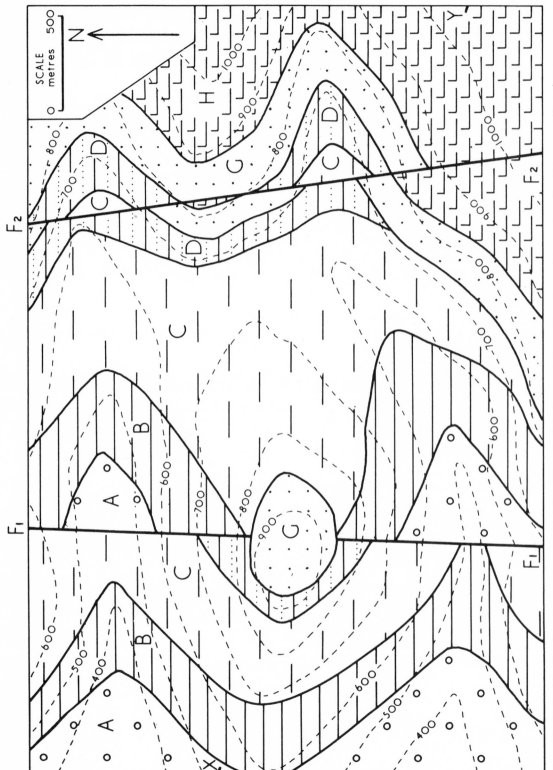

Map 16 Calculate the direction and amount of dip of the strata below and above the plane of unconformity. Draw a section along the line X–Y. The lines F_1–F_1 and F_2–F_2 are the outcrops of two fault planes. Which fault occurred earlier in geological time?

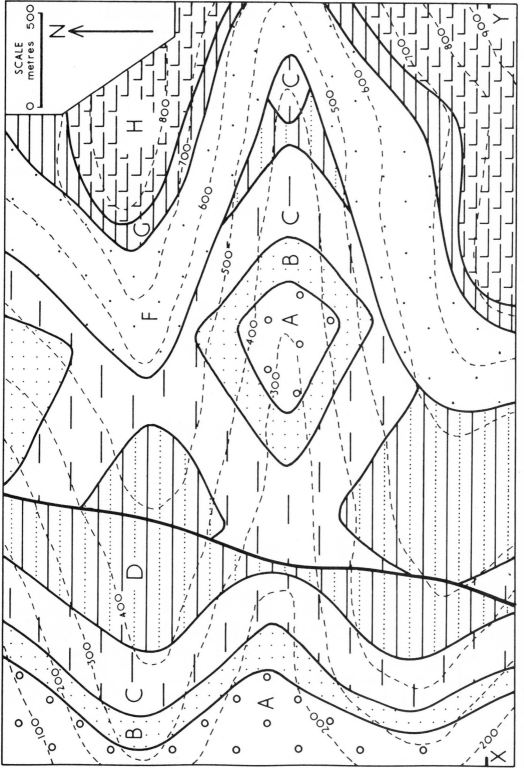

Map 17 This map includes all the structural features so far introduced; folds, a fault and an unconformity. Deduce the main structural features of the area from interpretation of the outcrop patterns before attempting to construct structure contours. Write a brief geological history of the area portrayed by the map, giving the order of events producing these structural features. Draw a section along the line X–Y.

39

6 More folds and faulted folds

Plunging folds

In the folds so far studied, the axis of the fold has been horizontal. The axis is the intersection of the axial plane with any bedding plane. (Make a synclinal fold by taking a piece of paper and folding it in two to make a simple V-shape. Drop a pencil into this and it will assume the position of the axis

called periclines and are a common occurrence.

Simple folds which have been subsequently tilted by further earth movements will also plunge. This is, furthermore, the simplest way of considering such structures although they may originate in other ways. This type of fold is referred

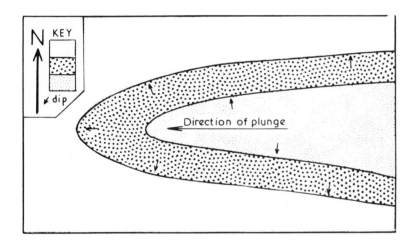

Fig. 30 Map showing the outcrops of anticlinally folded beds, plunging to the west.

of the fold.) Such a fold, with the axis horizontal, is called a *non–plunging* fold. The outcrops of the limbs tend to be parallel (but of course are affected by the configuration of the ground) since the structure contours drawn on one limb are parallel to those drawn on the other limb. (They have, as well, been parallel to the axial plane and, in cases where the axial plane was vertical, parallel to the axial trace (= outcrop of the axial plane).)

Where the fold axis is not horizontal, but is inclined, the fold is decribed as plunging. Some fold structures can be traced for many kilometres, others are of much more limited extent. Eventually a fold will die out by plunging (Fig. 30). Anticlinal folds which die out by plunging at both ends, giving rise to eliptical outcrop patterns, are

to in some earlier books as a 'pitching' fold, and not infrequently the terms 'pitch' and 'plunge' are used synonymously. The advanced student should consult F.C. Phillips's *The Use of Stereographical Projection in Structural Geology* 3rd edition (London, Edward Arnold, 1971), which deals fully with these terms. It is best to restrict the term 'pitch' to such linear features as striae, slickensides or lineations on a plane. They have both plunge (the angle they make with the horizontal) and pitch (the angle they make with the strike direction of the plane). Reject the use of the term pitch when referring to folds. Since problems entailing pitch can be solved (quite simply) by the use of stereographic projection they are not included in this book.

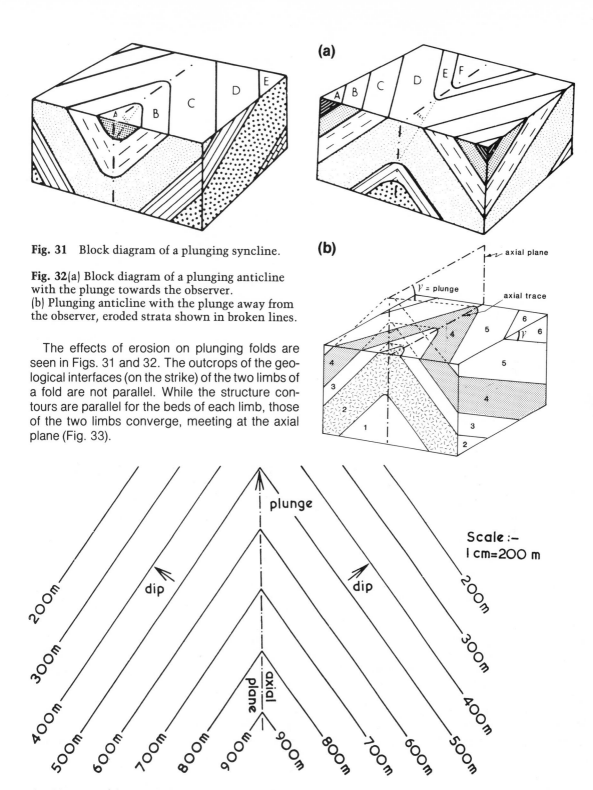

Fig. 31 Block diagram of a plunging syncline.

Fig. 32(a) Block diagram of a plunging anticline with the plunge towards the observer.
(b) Plunging anticline with the plunge away from the observer, eroded strata shown in broken lines.

The effects of erosion on plunging folds are seen in Figs. 31 and 32. The outcrops of the geological interfaces (on the strike) of the two limbs of a fold are not parallel. While the structure contours are parallel for the beds of each limb, those of the two limbs converge, meeting at the axial plane (Fig. 33).

Fig. 33 Map of the structure contour pattern of a plunging fold.

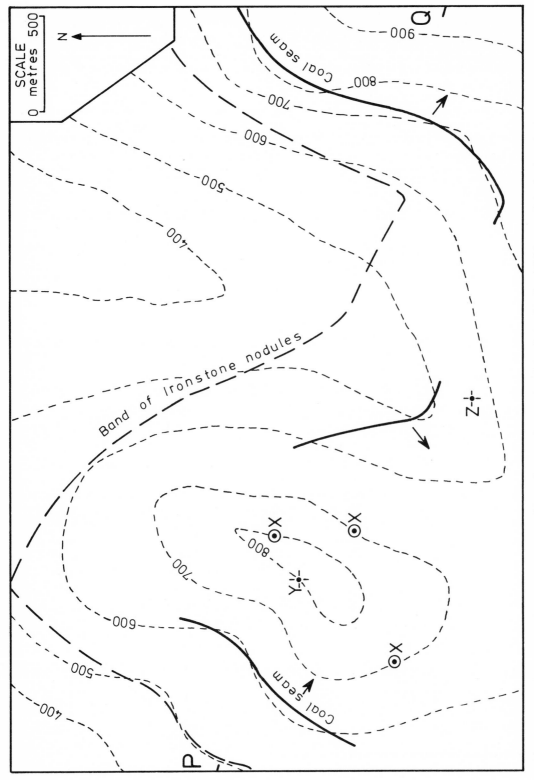

Map 18 The map shows part of the outcrop of a coal seam. It was also found at a depth of 300 metres in the boreholes at the several points marked X. Complete the outcrop of the coal seam. Determine the depth at which it would be encountered in shafts put down at Y and Z. Also insert the outcrop of another seam which lies 300 m higher (vertically) in the succession. What is the amount of plunge of the fold-axes? Insert outcrops of the fold axial planes on the map. Draw a section along the line P–Q.

42

Calculation of the amount of plunge Just as the inclination of the beds (dip) can be calculated from the spacing of the structure contours (see p. 6) measured in the direction of dip, so can the plunge of the fold be calculated from the spacing of the structure contours measured in the direction of plunge, i.e. along the axis. The plunge of the fold shown in Fig. 33 is, expressed as a gradient, 1 in 3.5 (to the north), if the scale of the diagram is 1 cm = 200 m, since the structure contour spacing measured along the axis is 1.75 cm. (The dip of the bedding plane of each limb is 1 in 2: check that this is so.)

Where we have no structure contours enabling accurate calculation of the angle of plunge, it is possible to give some estimate of it from width of outcrop. In Figure 30 the outcrop of the coarsely stippled bed is about twice as wide in the direction of plunge as it is in the limbs of the fold. We can conclude that the amount of plunge is considerably less than the angle of dip of the limbs of the fold. The relationship of dip to width of outcrop was discussed in Chapter 1.

The effects of faulting on fold structures

We have seen in Chapter 5 that the effect of dip-slip faulting (normal and reversed faults), followed by erosion, produces a lateral shift of outcrop in the direction of the dip of the strata on the upthrow side. Observe that on Figure 34 the vertical axial plane of a symmetrical fold is not displaced but the dipping planes show a shift of outcrop. Remember that the lower the angle of dip the greater the shift of outcrop.

Since the shift of outcrop is down-dip as a result of erosion, it follows that on the upthrow side of a fault the outcrops of the limbs of an anticline are more widely separated (than on the downthrow side). Conversely, the outcrops of the limbs of a syncline are closer together on the upthrow side (than on the downthrow) (Fig. 34).

On an asymmetrical fold, there will be a more pronounced shift of the outcrops of the limb with the lower angle of dip and a less pronounced shift of the outcrops of the steeper dipping limb. In the 'limiting' case of a monoclinal fold with a vertical limb there will be no shift of outcrops of the vertical beds. The axial plane of an asymmetrical fold dips, therefore the axial trace will be displaced.

In the case of overfolds – where both limbs dip in the same direction (see Fig. 16) – naturally the outcrops of both limbs will be shifted in the same direction by a dip-slip fault. Of course the outcrops of the shallow limb will be shifted more than those of the steeper dipping overturned limb.

Displacement of folds by strike-slip (wrench) faults

A wrench fault causes a lateral (horizontal) dislocation or displacement of the strata which may in some case be of many kilometres. The effect of a wrench fault is to displace the outcrops of beds, always in the same direction – and its effect on the outcrops of simply dipping strata in certain circumstances is similar to that of a normal fault: it may not be possible to recognize from the evidence a map can provide whether a fault has a vertical or a lateral displacement (see p. 33). However, the effects of a wrench fault on folded

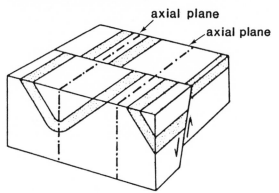

Fig. 34 Block diagram showing folded strata (one bed is stippled for clarity) displaced by a normal fault.

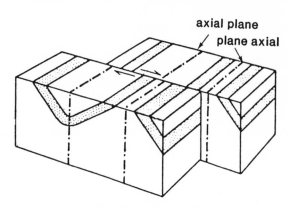

Fig. 35 Block diagram showing similar beds to Fig. 34 displaced by a wrench fault.

beds is immediately distinguishable from the effects of a normal fault since it will displace the outcrops of both limbs by an equal amount and in the same direction (whether the fold be symmetrical or asymmetrical) and, further, the outcrops of a bed occurring in the limbs of a fold will be the same distance apart on both sides of the fault plane (Fig. 35). The axial plane will, of course, be laterally displaced by the same amount as the outcrops of the beds occurring in the limbs of the fold, and this displacement occurs whether the axial plane is vertical (in a symmetrical fold) or inclined (in an asymmetrical fold). Compare Figs. 34 and 35.

Calculation of strike-slip displacement

Where any vertical phenomenon is present, for example the axial plane of a symmetrical fold, vertical strata, igneous dyke, etc. the lateral displacement by the fault can be measured directly from the map. If the strata are dipping and the fault is a purely strike-slip fault with no vertical component, or throw, (i.e. it is not an oblique-slip fault, see Fig. 25) then we can simply find the lateral shift by measuring the displacement of any chosen structure contour on a given geological boundary. Where strata are folded it is easy to measure the lateral displacement of fold axes or axial planes (Fig. 35).

Faults parallel to the limbs of a fold

So far we have considered faults which were perpendicular to the axial planes of the folds or, at least, cut across the axial plane. A fault which is parallel to the strike of the beds forming the limb of a fold is, of course, a strike fault. This will cause either repetition of the outcrops or suppression of the outcrops of part of the succession, in the manner discussed on p. 29.

Sub-surface structures

An interesting and important geological consideration is the deduction of the disposition of strata beneath an unconformity. In effect we are considering the 'outcrop' pattern on the plane of unconformity. If we could use a bulldozer to remove all

strata above that plane we would see the earlier, pre-unconformity strata outcropping.

This problem was touched on briefly in Map 8 where unconformity was introduced. If not already completed, return to Map 8 and insert the sub-unconformity outcrop of the coal seam. (Remember that topographic contours are now irrelevant; the surface we are considering is the plane of unconformity which is defined by the structure contours drawn on the base of Bed Y. Since both sets of structure contours are in each case straight, parallel and equally spaced – representing constant dips – their intersections giving the sub-Y outcrop of the coal seam should lie on a straight line.)

Posthumous folding

After the strata in an area have been laid down they may be uplifted, folded and eroded as we have seen in the last chapter. Further subsidence may cause the deposition of strata lying unconformably upon existing beds. Later still the processes of uplift and folding may recur, when this second period of folding is said to be posthumous – if the fold axis in the younger beds is parallel to, and approximately coincident with, the similar fold axis in the older beds. The trend of the two sets of folds may be parallel and the fold axes may coincide, as in Fig. 36, or they may be parallel but not coincident. An excellent example of this can be seen demonstrated by the Upper and Lower Carboniferous age beds in the Forest of Dean area included on the BGS 1:50,000 map of Monmouth (Sheet 233).

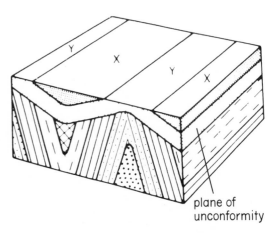

Fig. 36 Block diagram showing the effects of posthumous folding.

Since the folding is of two ages, the trends of the two sets of folds may be in quite different directions and it is then called superposed folding or cross-folding and is not included in posthumous folding, but there is usually a tendency for the earlier folding to exercise a 'control' over the later folding so that, more commonly, the trends are parallel or sub-parallel.

The strata of Pembrokeshire, see the Haverfordwest and Pembroke 1″ Geological Survey maps (Sheets 227, 228), were affected by Caledonian folding and later by Variscan (Hercynian) folding. In this area the trend of the folding of both periods is nearly east–west. By contrast, the rocks of the Lake District were gently folded with an east–west strike in Ordovician times and later by the Caledonian orogeny (phase in the mountain folding process) with a north-east–south-west strike, the later folding being of such intensity that it seems to have been independent of the control of the earlier folds.

Where the folds are of different trend it is possible to study the effects of one age of folding by taking a section parallel to the strike of the other folds. This was very neatly shown in the block diagram of a part of the southern Lake District in earlier editions of the 'Northern England' British Regional Geology.

Polyphase folding

Strata which have already been folded may at a later time be refolded. The stress causing the refolding may be due to a later phase of the same orogeny or even due to a later orogeny. The stress of the later phase of folding may be quite unrelated to the stress direction of the earlier folding (the 'first folds'). As a result of this the trends of the two ages of folding may be quite different.

Complex outcrop patterns are produced by the interference of two (or more) phases of folding. Simple examples are illustrated (Fig. 37). In some such simple cases the early folds can be seen to have been refolded since the axial planes of the first folds are themselves folded.

Bed isopachytes

All problems so far have dealt with beds of uniform thickness. However, traced laterally over some distance, strata may be seen to vary in thickness, a function of their mode of deposition. Such variations tend to be gradual and reasonably uniform within the area of a geological map sheet. Variations in the thickness of a bed are usually deduced from borehole data but may be discovered by measuring sections at geological outcrops and are occasionally revealed by variations in width of outcrop on a map.

The way in which a bed varies laterally in thickness can best be shown by constructing a series of bed isopachytes, lines joining points where the bed is known to be of the same thickness. To obtain the maximum number of control points at which the thickness of a bed can be determined it is usually necessary to construct two sets of structure contours, those for the top of the bed and those for its base. Their intersections give the thickness of the bed. Due to the nature of the sedimentary phenomena which produce beds of varying thickness, the isopachytes will tend to be reasonably straight (or gently curving) and approximately evenly spaced.

(a)

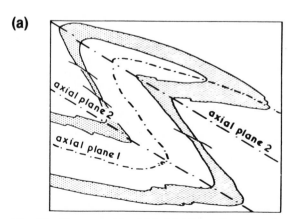

(b)

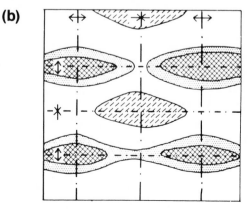

Fig. 37 Outcrops of refolded folds on flat ground: (a) a refolded isoclinal fold (b) second fold axes crossing first fold axes at right angles.

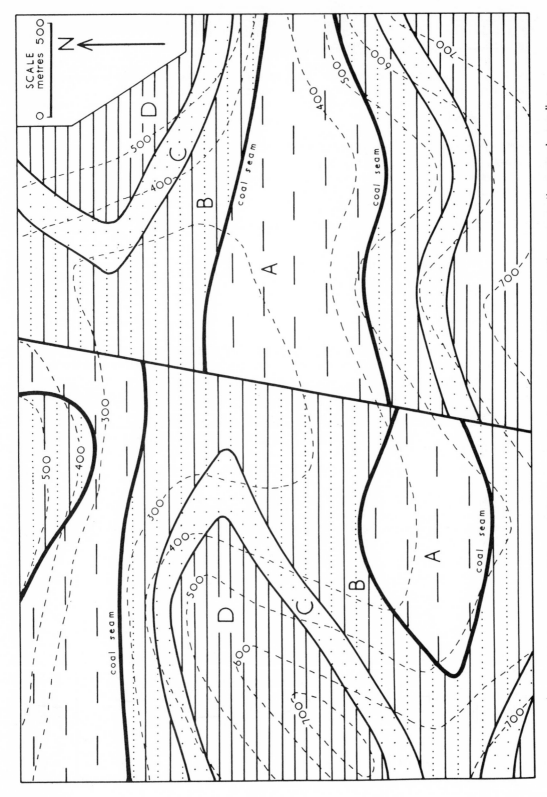

Map 19 What kind of fault occurs on this map? Has it any vertical displacement? Draw a section across the map along a line intersecting the fault. Indicate on the map anticlinal and synclinal axial traces.

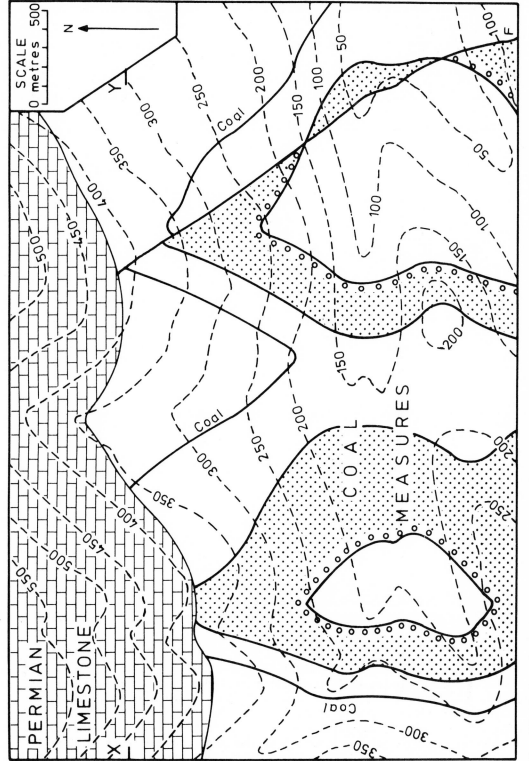

Map 20 Deduce the main structural features from the outcrop patterns. Insert the fold axial traces on the map. Draw a geological section along the line X–Y. Insert on the map the sub-Permian Limestone outcrops of the coal seam, the sandstone and the fault plane. To construct the latter it will be necessary to draw structure contours for the fault plane in order to deduce its dip.

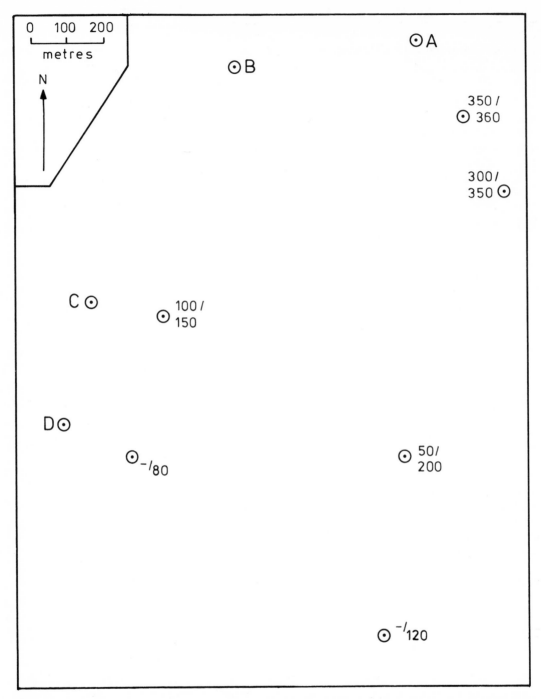

Map 21 A wedge-shaped, igneous, sill-like body is encountered in boreholes. The land surface is flat and horizontal over this area and the depths from the surface to the top and base of the sill are given beside each borehole. Construct structure contours for the top of the sill and its base — assuming that both top and bottom are uniformly dipping (but not parallel). Construct isopachytes at 50 m intervals to show how the sill varies in thickness. Shade the area where the sill outcrops. Why is the sill absent in boreholes A and B? If it was not found in boreholes at C and D how would you explain its absence?

Published Geological Survey Maps

Haverfordwest: 1:50,000 Map No. 228   Draw a section along the north–south grid line 197. Note the difference in amplitude of the folding in the Lower and Upper Palaeozoic rocks. Tabulate in chronological order the geological events.

Map 22   A good coal seam outcrops in a mountainous area of the Sierra Blanca, New Mexico. Shallow boreholes have been sunk at points along lines cut through dense pine forest to add to the meagre information derived from very old mine workings. It is proposed to mine the coal by opencast methods (stripmining). The data are here presented in simplified form: for each borehole three figures are given, namely, height of the ground above sea-level, height of the coal above sea-level, thickness of coal seam. All measurements are in feet. As in all consulting problems, rather more data would have been desirable.

   Construct a structure contour map for the coal seam using a 10 feet contour interval. Insert on the map the 50ft and 100ft overburden isopachytes. (These will pass through all points where the ground contours and the structure contours for the coal intersect and differ in height by 50 feet and by 100 feet. HINT: Naturally, the overburden isopachytes will be approximately parallel to the outcrop of the coal seam since in this quite small area of the map the topography is not complex. Draw a map to show the isopachytes for the coal seam thickness using a 'contour' interval of 0.5 feet (6.0, 6.5, 7.0, etc.) Note that 8.2 feet of coal are exposed in the mouth of the old coal adit. You may find it less confusing to draw this map on a tracing paper overlay. Note that the two pages of this map overlap at 'Join'. Either side can be folded and secured with a paperclip so that the join-lines coincide.

Map 23   The outcrops of the Middle Coal Measures (white) and the Lower Coal Measures (stippled) are shown. The heavy broken lines are the structure contours drawn on the base of the Lower Coal Measures (and have been deduced from borehole data). Construct structure contours for the Lower Coal Measures/Middle Coal Measures junction using outcrop information. (Note that the base of the Lower Coal Measures is folded, therefore the Lower/Middle Coal Measures boundary will also be folded about the same axial planes. Insert these on the map first). From intersections of the two sets of structure contours deduce the thickness of the Lower Coal Measures at as many points as possible. Draw isopachytes at 50 m intervals for the Lower Coal Measures. (Note that neither structure contours nor isopachytes can be drawn with a ruler in this example.) Drawing isopachytes can be done very conveniently on an overlay of tracing paper. Alternatively, different colours may be used for fold axial traces, structure contours and isopachytes. Draw a section along a north-west–south-east line using a vertical scale of 1 cm = 200 m (a vertical exaggeration of × 2.5).

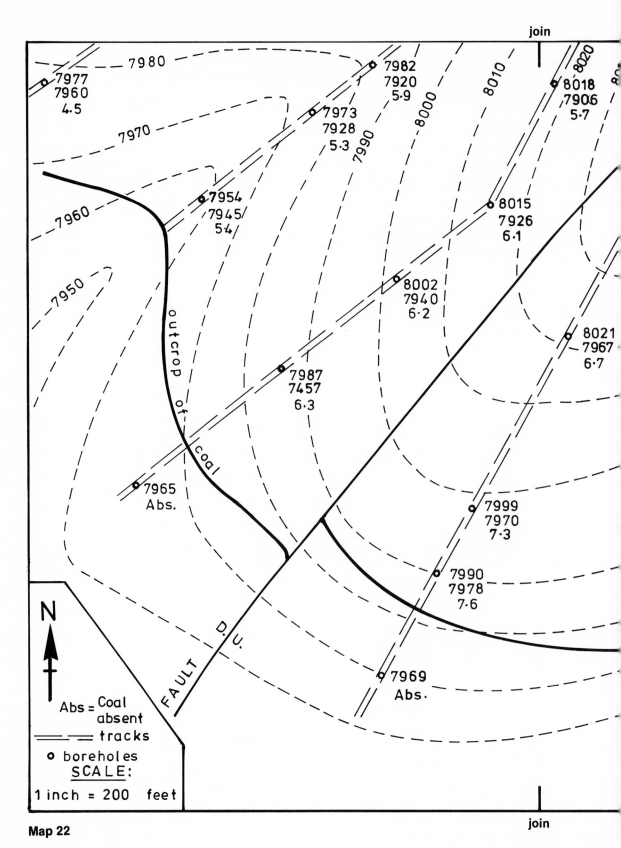

7980

7977
7960
4.5

7970

7982
7920
5.9

8010

8018
7906
5.7

8020

7960

7973
7928
5.3

7990

8000

7950

7954
7945
5.4

8015
7926
6.1

outcrop of coal

8002
7940
6.2

8021
7967
6.7

7987
7457
6.3

7965
Abs.

7999
7970
7.3

7990
7978
7.6

N

FAULT D. U.

7969
Abs.

Abs = Coal absent

tracks

o boreholes
SCALE:
1 inch = 200 feet

Map 22

50

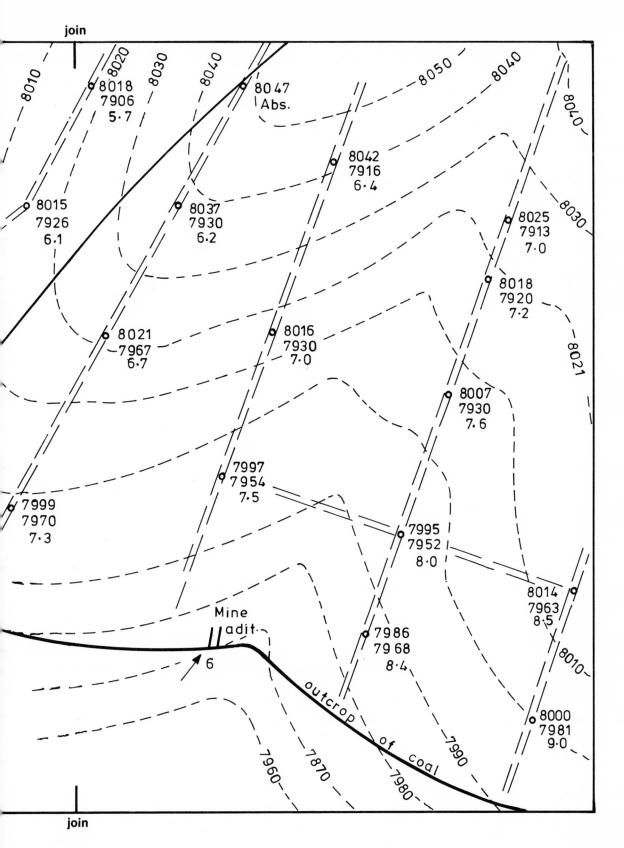

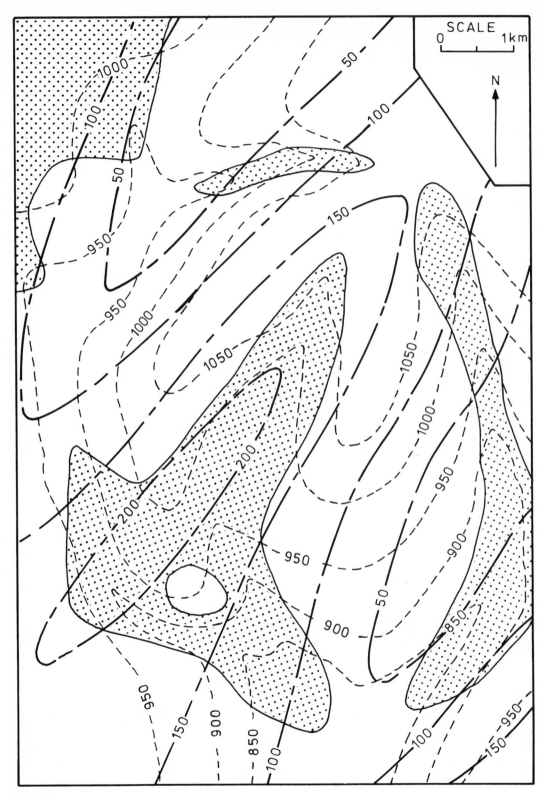

Map 23

7 Complex structures

Nappes

Following the discussion in Chapter 4 in which overfolds were described, structures in which beds are overturned beyond the vertical, we next consider nappe structures. A nappe arises from a very large overfold in which the strata are nearly horizontal over wide areas. The fold structure is referred to as a recumbent fold and has been 'pushed over' so far that both limbs have low angles of dip, and are approximately parallel, although in the case of one limb the beds are actually upside-down; i.e. the succession is inverted. Only at the 'nose' of the structure where the strata are folded back on themselves will steep dips be encountered (Fig. 38a).

Of course, minor folding is usually superimposed on a major fold such as this so that locally steep dips may be seen. This folding may be termed parasitic. The nature of the folding is related to its position on the overfold (Fig. 38b). This minor folding, frequently seen in the field, is of such a scale that it would be revealed only on large scale maps.

The axial plane of a recumbent overfold is nearly horizontal, but may be curved as in the above figure. Not infrequently the intense lateral stresses producing recumbent folds also cause rupture of the strata and produce a 'low angle' reversed fault (making a low angle with the horizontal). The thrust recumbent fold is a nappe structure (Fig. 39a).

Thrust faults

A thrust is a low angled fault plane along which movement has taken place, the strata above the thrust plane having been carried often for great distances in a near-horizontal direction, by intense earth movements, over the strata beneath. The thrust strata may have been displaced for a distance of many kilometres and may, for example, in the case of the Moinian rocks of Assynt (Sutherland) above the Moine Thrust, be quite different from any of the other rocks in the same district. On the other hand, the strata above the thrust may be similar to those beneath the thrust, but it should be noted that frequently older rocks may be thrust over younger ones. Thus it is possible to find Pre-Cambrian rocks overlying (due to thrusting) Cambrian rocks.

(a)

(b)

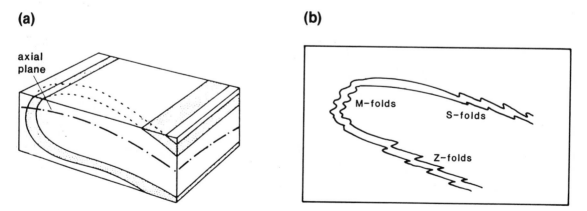

Fig. 38 (a) Idealized block diagram of a recumbent overfold. (b) The relationship of minor folds to their position on the overfold: M folds occur in the hinge region, S and Z folds are found on its limbs.

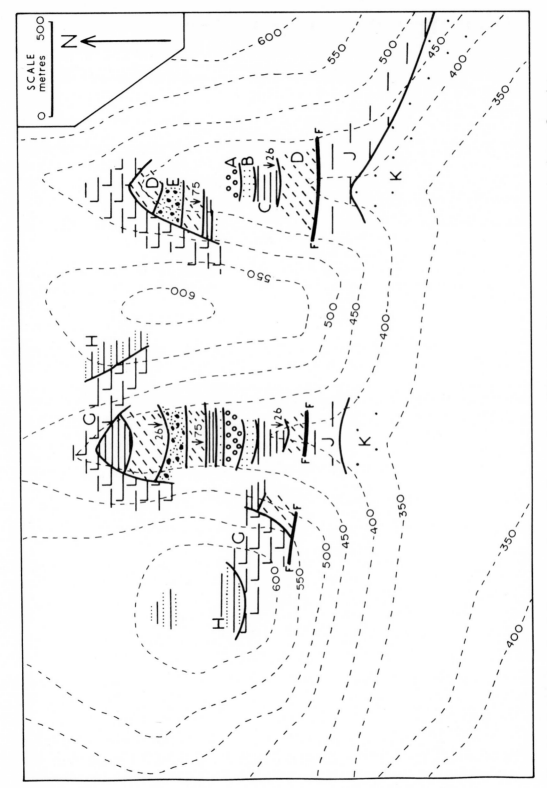

Map 24 Outcrops are few in the area portrayed by the map, but sufficient to enable structure contours to be drawn. Complete the outcrops over the whole map and draw a section along a north–south line. Briefly summarize the geological history.

The strata above a thrust plane may be approximately parallel to it but, on the other hand, their dips may be unrelated to the inclination of the thrust plane, a whole block of rocks having been moved *en masse*. Frequently, just above a thrust plane, dips are locally affected by the movement along the thrust, beds being overturned due to the effects of drag over the rocks beneath the thrust.

The strata beneath a thrust plane may be very greatly affected by the forces associated with the thrusting. The effect on these beds is to produce imbricate structures. These comprise many parallel or near-parallel faults, sometimes of high angle of dip although they are reversed faults, which divide the unthrust area or foreland into 'slices' (Fig. 39b).

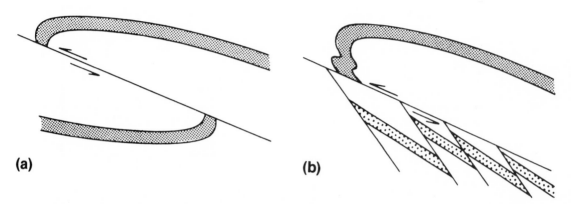

(a)　　　　　　　　　　　　　**(b)**

Fig. 39 (a) Section of a recumbent fold which has been thrust (a nappe) and (b) section showing imbricate structures which commonly occur beneath such a thrust plane.

Axial plane cleavage

When rocks which are bedded (possessing what may be termed primary structures) are subjected to pressures – or stresses – they may, as we have discussed, become fractured by faults or they may become crumpled into fold structures. The stress applied to a rock may also cause it to be deformed and new structures are formed such as cleavage and schistosity. Cleavage is developed as a result of shortening of the rock in a direction perpendicular to the cleavage planes with a stretching or extension of the rock in the plane of the cleavage.

Cleavage is often found in rocks which are folded. The greatest shortening of strata due to folding is perpendicular to the fold axial planes

and it therefore follows that cleavage planes are parallel to the fold axial planes, hence the cleavage is called axial plane cleavage. (This will not be true for a sequence of beds of different competency nor, of course, in the case of complex refolded folds.)

Cleavage dips, indicated by a symbol such as ⟍, should not be confused with dips of bedding planes – to which there may not seem at first sight to be an obvious relationship. This information must not, of course, be dismissed as something clouding the issue: if the cleavage is axial plane cleavage (the most frequently found) there will be a consistent relationship between the cleavage direction and the main structural features (folds). The cleavage will be parallel to the axial planes of the folds.

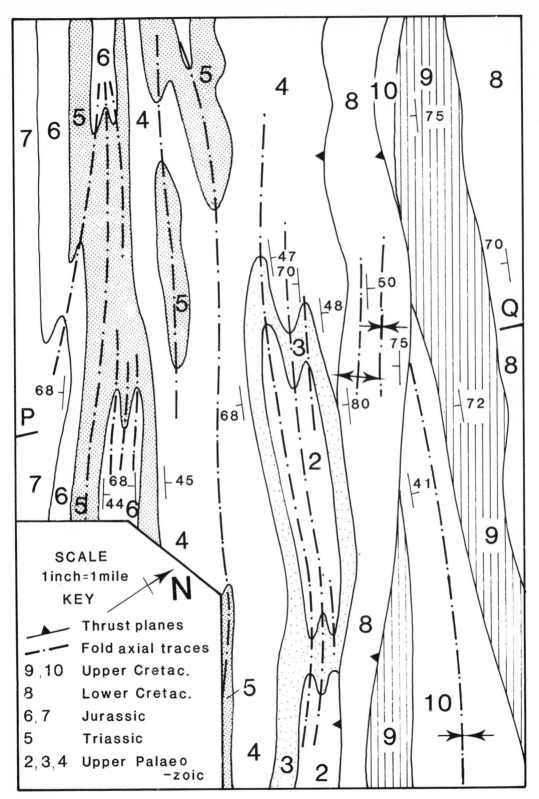

7   6   5 ᴎ   4   6   5   4   10 8   9   8

68

P

7
6   5   44   6   68   45
4

SCALE
1inch=1mile
KEY

N

Thrust planes
Fold axial traces
9,10   Upper Cretac.
8   Lower Cretac.
6,7   Jurassic
5   Triassic
2,3,4   Upper Palaeo
-zoic

47
70
48
50
3
75
80
68
72
2
41
Q
8
9

8

5
10

4   3   2   9

Map 25

56

The relationship between the cleavage dip and the dip of the bedding reveals, in overfolds, in which limb the beds are the right way up and in which limb the beds are inverted. Cleavage dip steeper than bedding = right way up: bedding dip steeper than cleavage dip = inverted (Fig. 40). This 'rule' is necessarily true when there has been only one period of folding.

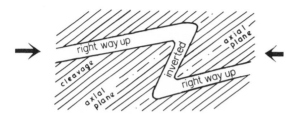

Fig. 40 Section showing cleavage/bedding relationships in overfolds with axial plane cleavage.

Published Geological Survey Maps
Assynt: 1″ Geological Survey map (Special Sheet) How may the thrust planes be distinguished from the unconformities? Find on the map some examples of imbricate structures.

Map 25 is adapted from a portion of the 1 inch to the mile sheet 83 $\frac{E}{10}$ Adams Lookout, Alberta, Canada. It is a highly folded area of Upper Palaeozoic and Mesozoic strata. Some fold structures persist longitudinally for many miles with a roughly NW–SE trend (Note the direction of the North point on this map). Other plunging folds, especially minor folds, are only locally developed. There are steeply dipping thrust faults (= reversed faults). They are strike faults with a south-westerly dip of about 50°. In reality relief is considerable, 1200 feet from valley bottom to peaks, but contours have been omitted from Map 25 for simplicity. Draw a cross-section (true scale) along the line P–Q on the profile provided (Fig. 41). Most of the fold axial traces are shown on the map. Indicate any others which may have been omitted. Show which folds are synclinal (mark with S) and which are anticlinal (mark with A).

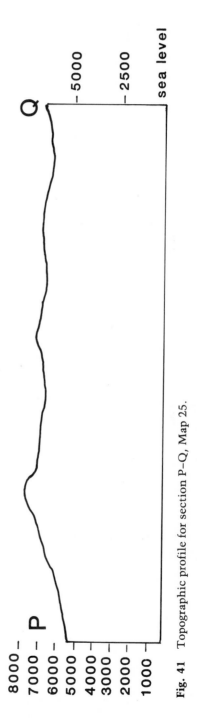

Fig. 41 Topographic profile for section P–Q, Map 25.

8   Igneous rocks

Igneous rocks may be divided, according to their mode of occurrence, into extrusive (lava poured out onto the surface) and intrusive rocks (injected into existing rock). The intrusive igneous rocks may be further classified according to their form, but they fall from a structural point of view into two groups, namely, transgressive or discordant intrusions and concordant intrusions.

Concordant intrusions

Sills   The chief intrusions of this category are sills, 'beds' ranging in thickness from metres to a few millimetres of igneous material, commonly dolerite or basalt or felsite, which are parallel to the bedding of the sedimentary rocks into which they have been intruded. The intrusion seldom causes any observable disturbance of strata so that a sill behaves structurally as though it were part of the stratigraphical succession: if, subsequent to intrusion, the beds are tilted or folded then the sill is also tilted with the same dip or is folded.

One respect in which a sill is distinguishable from the strata including it, apart from its petrology, is that it is younger than those strata: it has been intruded after they were laid down and consolidated, perhaps long after the sedimentary beds were formed. Consequently, the strata into which a sill is intruded may already have been faulted, so that we may find beds displaced by faults which do not displace the sill. Of course, faulting which occurs after the intrusion of a sill will displace sill and strata alike.

A sill although normally concordant, may change its horizon occurring at one level in the sequence at one locality but at a different horizon at another locality. Often such changes in horizon or stratigraphical position are accomplished in 'steps', i.e. the changes are abrupt, the molten material being intruded taking advantage of lines of weakness such as joints or existing fault planes (Fig. 42).

Sills may also show changes in thickness, for example the Great Whin Sill of northern England may be as much as 39 metres thick in some outcrops but in others less than 30 metres. A sill may split laterally into 'leaves', a phenomenon not seen in lava flows. It is possible, however, to confuse this with rafts of sedimentary strata sometimes caught up in lava flows.

Sills, being composed of resistant rock, may often be found capping hills, for example those on the Fife–Kinross border. A sill may form a pronounced escarpment such as the one on which Stirling Castle is built or the one which runs across Northumberland (made by the Great Whin Sill) on which part of Hadrian's Wall is built.

Other concordant intrusions such as laccoliths and lopoliths, the form of which are described and defined in any petrology book, are of much less frequent occurrence and are not dealt with here.

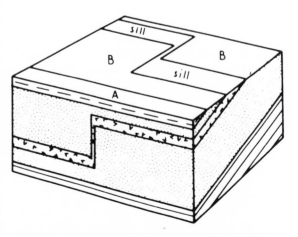

Fig. 42   Block diagram showing a sill intruded into dipping strata. Note that the sill is seen to change horizon abruptly.

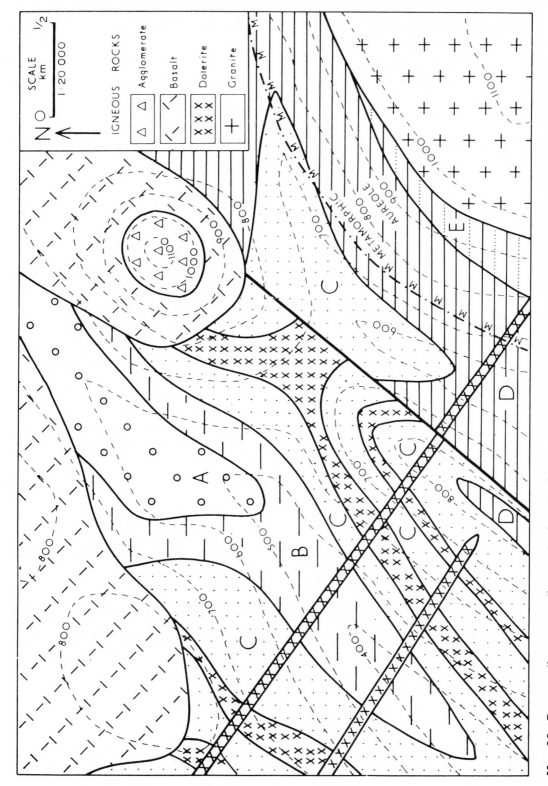

Map 26 Draw sections across the map to show the form of the igneous rocks and the other structural features. Also deduce the relative ages of the igneous rocks as far as this is possible.

Lava flows and tuffs

These are the products ejected from active volcanoes, but in Britain there are many examples resulting from former periods of vulcanicity. From map information alone a lava flow may be difficult to distinguish from a sill, especially as both may be classed petrologically as basalt: rock type is not a reliable criterion. If a lava is poured out on to an eroded surface it may come to rest on beds of different age, producing in effect an unconformity. If it is poured out on the sea bed or on to recently uplifted un-eroded sediments then it will be 'conformable' and structurally resemble a still. The field evidence that the upper surface of a lava is frequently weathered whereas that of a sill is not, and the fact that a sill thermally metamorphoses the overlying sediments, cannot be deduced from map evidence.

Volcanic ash which falls into water will be mixed with detrital sediment which is being deposited giving rise to ashy sediment. When volcanic activity is more intense beds will be comprised entirely of volcanic ash. Ash beds may also be deposited on land. Beds of consolidated ash, known as tuff, may be interbedded with lavas or with sedimentary strata and, in the latter case may be regarded as sedimentary rocks despite their volcanic origin.

Discordant intrusions

Dykes These are vertical or near vertical intrusions, commonly of dolerite or basalt, varying from a few centimetres in width to several metres (but rare examples are tens of metres wide). They are frequently traceable across country for many kilometres, cutting through the strata into which they have been intruded.

Since they were approximately vertical when intruded their outcrops are usually straight, regardless of variations in topography. It is pos-

sible, in certain cases, to date a dyke relative to the sedimentary strata: it must be later in origin than the youngest beds which it cuts and, in addition, where an unconformable series of younger strata occur it is possible to see whether a dyke cuts both the younger and older series (cf. the dating of faults, p. 35).

Dykes are intruded during periods of crustal tension. Widening cracks or joints permit igneous material to be intruded from below (and laterally). They may be very numerous near to volcanic centres where they may represent a considerable extension of the crust, for example see the south coast of the Isle of Arran (Map 27).

Ring-dykes A special example of dykes, circular in plan or arcuate with a diameter up to 5 kilometres, ring dykes are vertical or very steeply dipping. In some cases they are associated with down-faulting of rocks within the ring.

Cone-sheets Also circular in plan, but seldom completing the circle, cone-sheets dip steeply inwards towards a centre. They tend to occur in 'swarms'. They are usually of basaltic material, whereas ring-dykes are more commonly granitic. Cone-sheets are seldom more than a metre or two thick but ring-dykes may be tens of metres thick. Examples of both can be seen on BGS maps covering parts of the Hebridean Tertiary Volcanic Province.

Stocks, bosses and batholiths These are larger bodies of intruded material. Of plutonic origin, intruded deep into the earth's crust and now exposed at the surface only through prolonged erosion, they are most commonly of granite. Varying in plan from near circular (bosses) to irregularly shaped masses (stocks) they also vary greatly in size, the larger intrusions (batholiths)

Map 27 This map is broadly adapted from the geological map of the Isle of Arran 1:50,000 Special Sheet. Most of the kinds of igneous intrusion shown on the stylized Map 26 can be found here. Write an account of the geology of the area, noting particularly the two major unconformities, the dips of strata of different ages and the distribution and disposition of different igneous rocks. What can you deduce from the general direction of the dykes? It is possible to work out the geological history for this map (however, it will not be identical to that of the one-inch to the mile geological survey sheet). Reproduced by permission of the Director, British Geological Survey: NERC copyright reserved.

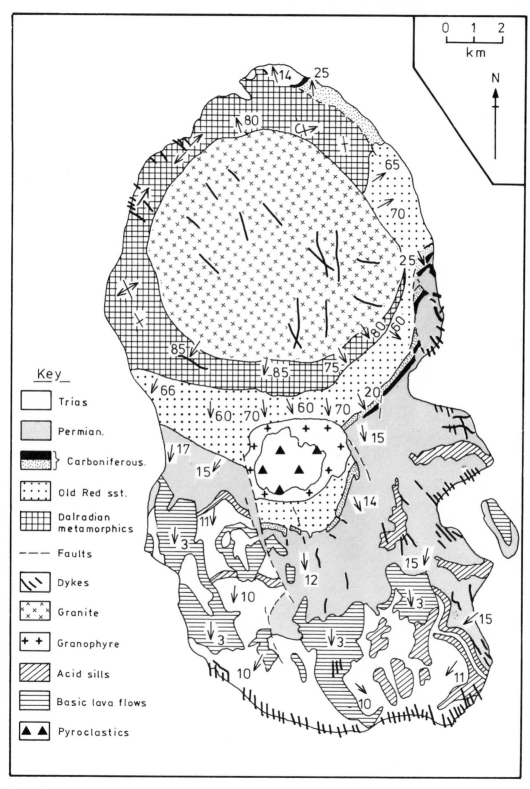

Key

- ☐ Trias
- ▦ (shaded) Permian.
- ■ }Carboniferous.
- ⋮ Old Red sst.
- ▦ Dalradian metamorphics
- – – – Faults
- ⧄ Dykes
- ⊠ Granite
- + + Granophyre
- ⧄ Acid sills
- ≡ Basic lava flows
- ▲ ▲ Pyroclastics

Map 27

being, exceptionally, up to hundreds of kilometres in length. Often these discordant intrusions are longer than they are broad, being elongated parallel to the general strike of folding and other structural features, but in every case their margins are vertical or steeply dipping, cutting discordantly across the sedimentary beds. Made of resistant rock, they commonly form high ground. Areas where the sediments have been baked and altered by the heat from the intrusion occur surrounding the larger igneous intrusions. These are known as metamorphic aureoles.

Volcanic necks   These are the infilling of volcanic vents with consolidated lava (basalt, etc.) or pyroclastic material (agglomerate or tuff). They cut through existing strata and have vertical or near vertical sides and they are usually nearly circular in plan. A volcanic neck structurally, therefore, resembles a boss, but is normally of much smaller diameter and the rock types encountered are entirely different.

It is sometimes difficult from map evidence alone to distinguish between a basalt capping a rounded hill (a sill or a lava flow), possessing a circular outcrop, from a volcanic neck with a basalt infilling also having a circular outcrop, although in section the two features would be very different (Fig. 43).

Maps of areas of more complex structure and with a degree of metamorphism often include information on cleavage (see p. 55). Note that cleavage dips are given on Map 29 and they provide a useful clue in solving the structural problems.

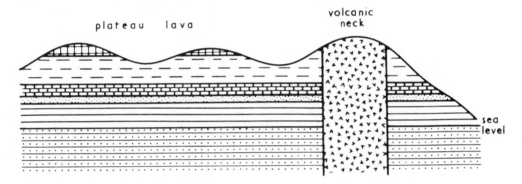

Fig. 43   Geological section showing a volcanic neck and lava-capped hills.

Published Geological Survey Maps
Edinburgh: 1″ Map No. 32 (Scotland) Give the relative ages of the various types of igneous intrusion shown on the map. How can the relative age of an igneous rock be discovered from map evidence and what are the limitations of this method? What other evidence might one expect to find in the field but which cannot be discovered from a map? (This map has now been replaced by 32E Edinburgh and 32W Livingstone on the 1:50 000 scale.)

Map 28   This is part of a geological map produced in the field (in northern England). Actual outcrops and exposure of geological boundaries are limited because of vegetation and the presence of superficial deposits of peat, alluvium, etc. not shown on the map. Complete the geological outcrops on the map. Draw a section along a north–south line to illustrate the structures. Write a geological history of the area including notes on the relative ages of the igneous rocks.

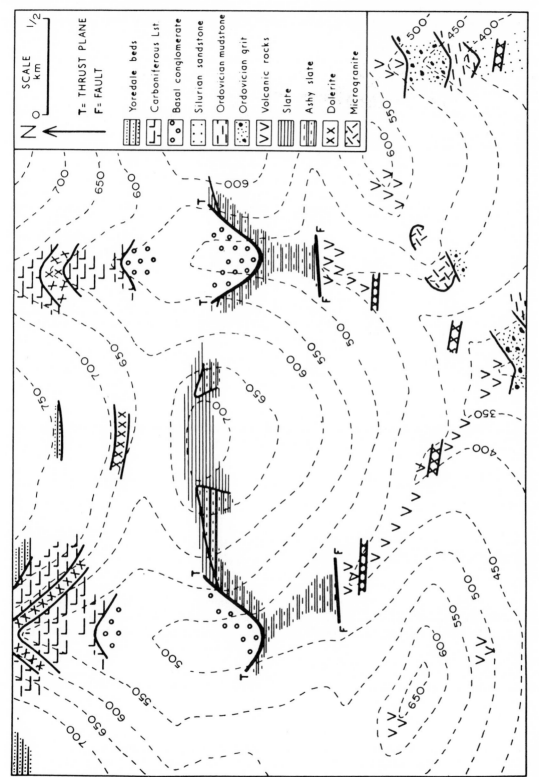

Map 28

Description of a geological map

All sources of information available should be used and coordinated, the information deducible from the map itself, the column of strata usually provided and sections showing the geological structures (the latter if not given on the map, drawn by the student). The description of a map should summarize the strata present. The general structural pattern and the trend of the chief structural features should be deduced from outcrop patterns. The broad relationship of topography to geology should be noted. It is usual to summarize the economic geology. Most important, a geological history of the area should be described. This comprises an attempt to relatively date all geological events, the deposition of sediments, breaks in the succession (unconformities), the intrusion of igneous rocks and the development of faults and folds. The evolution of superficial deposits, the drainage pattern and topography complete the history. In some cases a complete chronology cannot be deduced from map evidence and some events cannot be dated but their possible age and the ambiguity of the evidence must be discussed. Often it is useful to illustrate your answer with a structural sketch map. An example is given of the main deductions which may be made from Map 28.

The geological history of Map 28

The earliest bed present is the ashy slate – formed by tuff falling into a sea in which argillaceous sediment was being deposited. A lull in vulcanicity but with continued deposition accounts for the overlying strata, now slate. The age of both formations is in doubt: they are thrust over Carboniferous Basal Conglomerate (which is clearly younger) and they are faulted against volcanic rocks. Though metamorphism itself is no guide to antiquity of strata, these slates (and ashy slates) may be the oldest strata present, since no other beds have been metamorphosed, and thus older than the volcanics which are

earlier than the Ordovician grit. After the volcanic episode, perhaps prolonged although the thickness of volcanic rocks is not deducible since their structure is unknown, a conformable sequence of sediments was laid down. This comprises Ordovician grit and mudstone and Silurian sandstone. The lowest of these appears to rest partly on the microgranite which must therefore have been intruded at an earlier date. The age of the east–west fault must be later than the slates and the volcanics which it cuts, but how much later cannot be established.

Dolerite dykes cut volcanic rocks and beds as young as the Silurian sandstone, and hence were intruded after the deposition of these beds. Although not cutting Carboniferous rocks the dykes *may* be contemporaneous with the post-Carboniferous sill, also of dolerite.

After (probably prolonged) non-deposition (since no Devonian age strata are seen) Carboniferous seas spread into the area depositing initially a basal conglomerate, though its base is not seen and we do not know what it rests on. As the sea became clearer (?deepened) limestone was deposited, followed by the Yoredale beds – cyclic sediments laid down in shallow marine to terrestrial conditions. Two major post-Carboniferous events occurred: the thrusting northwards of older rocks over the Carboniferous, accompanied by overfolding of strata above the thrust plane, and the intrusion of the thick dolerite sill. Neither can be dated precisely. Both may be approximately the same age, referable to the late Carboniferous orogeny (the Variscan) although either could be of much later date. A northerly tilt was imparted to the area since the horizontally deposited Carboniferous strata have a northerly dip.

Mesozoic and Tertiary events are unknown since no strata of this age occur here. However, uplift and subsequent erosion have given rise to the present topography, southwards sloping valleys cutting back into a high east–west escarpment. The topography relates closely to the underlying geology. Superficial deposits are not shown on this map.

Map 29 Describe the geological structure of the area of the map and draw a section along the line X–Y.

65

Numerical Answers

Map 1
Sandstone 2   100 m
Mudstone   150 m
Shale   50 m
Sandstone 1   150 m

Map 3
1 in 2½; Bearing 174°
(6° E. of South)

Map 4
1 in 5; East
B   250 m
C   100 m
D   100 m
E   50 m

Map 5
200 m

Map 8
A   450 m
B   200 m
C   absent

Map 12
500 m
200 m

Map 18
Y   400 m
Z   110 m
1 in 4 (14°)

Map 19
Wrench fault
No throw
Lateral displacement
3.2 cm = 615 m.

Table I

If vertical thickness
of a bed (V.T.) = 100 m

Angle of Dip:	True Thickness of bed:
0	100
10	98.5
20	94.0
40	76.6
60	50.0
80	17.4

Table II

True Dip:	If angle between line of section and the direction of True Dip is:				
	10°	30°	50°	70°	90°
	then Apparent Dip in section is:				
0	0	0	0	0	0
10	10	9	6	3	0
30	30	27	20	11	0
50	50	46	37	22	0
70	70	67	60	43	0
90	90	90	90	90	—

Apparent dips given to the nearest degree

Index

AAPG, Tulsa, 10
Adams Lookout, Alberta, 57
angle of dip, 3
angular unconformity, 16
anticline, 19
apparent dip, 6
Arran, Isle of, 60
Assynt, 18, 53, 57
asymmetrical fold, 20, 43
aureole, metamorphic, 62
axial plane, 19
  cleavage, 55
axis, 40
Aylesbury, 10

batholith, 60
bed isopachyte, 35, 45
borehole, 11, 12, 13, 37
  depth in, 13, 16
boss, 60
boundary, geological, 1, 5, 13
Brighton, 19
British Geological Survey, 3, 10
buried landscapes, 18

Chester, 31
Chesterfield, 37
cleavage, 55
  dip, 55
classification of faults, 28
compass bearing, 5
  Brunton, 5
competent rocks, 23
concordant intrusions, 58
cone-sheet, 60
contour, 1
Cotswold Hills, 1

depth in borehole, 11, 16
derived fossils, 16
  fragments, 16
dip, 3, 6, 27
  apparent, 6
  fault, 29, 31
  in cuttings, 6
  reversed, 21

  true, 6
dip-slip fault, 28
discordant intrusions, 60
displacement of outcrops, 31, 43
downthrow, 27
dyke, 60

Edinburgh, 62
extrusive igneous rocks, 58

fan structure, 21
fault, 27
  dating of, 35
  'hade', 27
  heave, 27, 33
  line scarp, 28
  plane, 27
  plane dip, 27
  scarp, 28
  throw, 27, 31
faulting, effects on outcrop, 28
  posthumous, 35
faults, 27
  classification of, 28
  dip, 29, 31
  dip-slip, 28, 31
  normal, 27
  oblique, 29
  oblique-slip, 28
  reversed, 27
  strike, 29, 31, 44
  strike-slip, 28, 33, 44
  tear, 33
  thrust, 55
  wrench, 33, 43
first folds, 45
fold, 19
  axial plane, 19
  axis, 40
  limb of, 19
  plunge, 40, 43
folding, 19
  minor, 19, 53
  parasitic, 53
  polyphase, 45
  posthumous, 44

superposed, 45
folds, 19
    anticline, 19
    asymmetrical, 20, 43
    close, 21
    concentric, 21
    fan, 21
    flat, 20, 21
    gentle, 21
    inclined, 20
    isoclinal, 20, 21
    non-plunging, 40
    open, 21
    plunging, 40
    posthumous, 45
    recumbent, 53
    refolded, 45
    similar, 21
    straight-limbed, 23
    symmetrical, 19
    syncline, 19
    upright, 19
foreland, 55
Forest of Dean, 44
fossils, derived, 16
fragments, derived, 16

geological boundary, 1, 5, 13
    history, 39, 64
    interface, 5
Geological Highway Map, 10
Geological Survey maps, 3, 10, 18, 19, 25, 28,
    31, 37, 44, 45, 49, 57, 60, 62
gradient, 6
Great Whin Sill, 58

'hade', 27
Hadrian's Wall, 58
hardgrounds, 16
Haverfordwest, 45, 49
heave, 27, 33
Hebridean Volcanic Province, 60
Henley-on-Thames, 10
horizontal strata, 1

igneous rocks, 58
    extrusive, 58
    intrusive, 58
imbricate structure, 55
inclined strata, 3
incompetent rocks, 23
inlier, 9, 35
insertion of outcrops, 13

inter-limb angle, 19
intrusions, concordant, 58
    discordant, 60
inverted strata, 21, 53, 57
isoclinal fold, 20, 21
isopachyte, 35
    bed, 35, 45
    overburden, 35

laccoliths, 58
Lake District, 45
lava flow, 60
Leeds, 37
Lewes, 25
limb of a fold, 19
lopliths, 58

marine erosion, plane of, 18
M-folds, 53
Moinian, 55
Monmouth, 44
Moreton-in-the-Marsh, 3

nappe, 53
non-plunging fold, 40
normal fault, 27
numerical answers, 66

off-lap, 16
outlier, 9, 35
overburden, 35
    isopachyte, 35
overfold, 20, 53, 55
overlap, 16
overstep, 16
overturned beds, 21

paleosols, 16
Pembroke, 45
pericline, 40
Phillips, F.C., 40
'pitch', 40
plane of unconformity, 16
plunge, 40, 43
plunging fold, 40
polyphase folding, 45
posthumous faulting, 35
    folding, 44
projection, stereographic, 40

recumbent fold, 53
repetition of outcrops, 20, 29
Representative Fraction, 3

reverse fault, 27
reversed dip, 21
ring dyke, 60

scale, 6
section drawing, 1, 6
S-folds, 53
Shrewsbury, 18
Sierra Blanca, N.M., 49
sill, 58
stereographic projection, 40
stock, 60
strata, 1
stress, 23, 27
strike, 3, 5
    fault, 29, 31
    line, 5
strike-slip fault, 28, 33, 44
structure contour, 5, 11, 13, 28, 35
sub-surface structures, 44
sub-unconformity outcrops, 18, 44
superposed folding, 45
symmetrical fold, 19
syncline, 19

tear fault, 33

thickness of a bed, 8
three-point problem, 11
throw of a fault, 27, 31
thrust fault, 55
    plane, 55
topographic profile, 1
true dip, 6
tuff, 60

unconformity, 16, 35
    plane of, 16
upthrow, 27

vertical beds, 9, 43
    exaggeration, 3, 10, 25
    scale, 1, 3
    thickness, 8
volcanic ash, 60
    neck, 62

'way-up', 57
width of outcrops, 9, 20, 43
wrench fault, 33, 43

Z-folds, 53